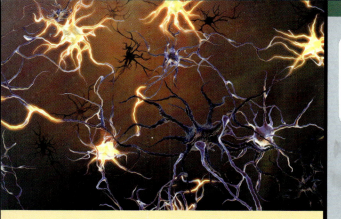

The Benjamin Cummings Editorial Team

Author, end-of-article material:
Jay L. Brewster
Associate Professor of Biology
Pepperdine University

Selections Editor:
Susan Winslow
Acquisitions Editor, Biology
Benjamin Cummings Science Publishers

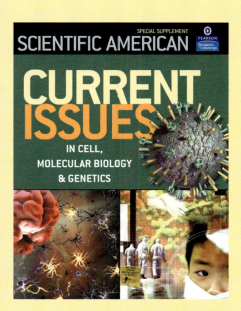

ON THE COVER
top: George Retseck
bottom left: Jeff Johnson/*Hybrid Medical Animation*
bottom right: Photoillustration by Jason Jaroslav Cook;
NIBSC/PHOTO RESEARCHERS, INC. (*virus SEM*);
Vincent Yu / *AP Photo* (*line of people*);
Hoang Dinh Nam / AFP / GETTY IMAGES
(*boy with mask and doctor with x-ray*)

MW01519407

Current Issues in Cell, Molecular Biology & Genetics is published by
Scientific American, Inc. with project management by:

Jeremy Abbate
PROJECT DIRECTOR, MANAGER OF CUSTOM PUBLISHING

Silvia De Santis
DESIGNER, CUSTOM CONTENT

The contents of this issue are adaptations of material
previously published in SCIENTIFIC AMERICAN.

EDITOR IN CHIEF: John Rennie
EXECUTIVE EDITOR: Mariette DiChristina
MANAGING EDITOR: Ricki L. Rusting
NEWS EDITOR: Philip M. Yam
SPECIAL PROJECTS EDITOR: Gary Stix
SENIOR EDITOR: Michelle Press
SENIOR WRITER: W. Wayt Gibbs
EDITORS: Mark Alpert, Steven Ashley, Graham P. Collins, Steve Mirsky, George Musser, Christine Soares

ART DIRECTOR: Edward Bell
SENIOR ASSOCIATE ART DIRECTOR: Jana Brenning
ASSOCIATE ART DIRECTOR: Mark Clemens
ASSISTANT ART DIRECTOR: Johnny Johnson
PHOTOGRAPHY EDITOR: Emily Harrison
PRODUCTION EDITOR: Richard Hunt

COPY DIRECTOR: Maria-Christina Keller
COPY CHIEF: Molly K. Frances
COPY AND RESEARCH: Daniel C. Schlenoff, Michael Battaglia

EDITORIAL ADMINISTRATOR: Jacob Lasky
SENIOR SECRETARY: Maya Harty

ASSOCIATE PUBLISHER, PRODUCTION: William Sherman
PREPRESS AND QUALITY MANAGER: Silvia De Santis
PRODUCTION MANAGER: Christina Hippeli

DIRECTOR, ANCILLARY PRODUCTS: Diane McGarvey

ASSOCIATE PUBLISHER/VICE PRESIDENT, CIRCULATION: Lorraine Leib Terlecki

GENERAL MANAGER: Michael Florek
BUSINESS MANAGER: Marie Maher

CHAIRMAN: John Sargent

PRESIDENT AND CHIEF EXECUTIVE OFFICER: Gretchen G. Teichgraeber

VICE PRESIDENT: Frances Newburg

Subscription inquiries for SCIENTIFIC AMERICAN magazine:
U.S. and Canada (800) 333-1199; other (515) 247-7631, or www.sciam.com.

To learn more about Scientific American's
Custom Publishing Program, call 212-451-8859.

New Movement in PARKINSON'S

Recent genetic and cellular discoveries are among the advances pointing to improved treatments for this increasingly common disorder

By Andres M. Lozano and Suneil K. Kalia

Parkinson's disease, first described in the early 1800s

by British physician James Parkinson as "shaking palsy," is among the most prevalent neurological disorders. According to the United Nations, at least four million people worldwide have it; in North America, estimates run from 500,000 to one million, with about 50,000 diagnosed every year. These figures are expected to double by 2040 as the world's elderly population grows; indeed, Parkinson's and other neurodegenerative illnesses common in the elderly (such as Alzheimer's and amyotrophic lateral sclerosis) are on their way to overtaking cancer as a leading cause of death. But the disease is not entirely one of the aged: 50 percent of patients acquire it after age 60; the other half are affected before then. Furthermore, better diagnosis has made experts increasingly aware that the disorder can attack those younger than 40.

So far researchers and clinicians have found no way to slow, stop or prevent Parkinson's. Although treatments do exist—including drugs and deep-brain stimulation—these therapies alleviate symptoms, not causes. In recent years, however, several promising developments have occurred. In particular, investigators who study the role proteins play have linked miscreant proteins to genetic underpinnings of the disease. Such findings are feeding optimism that fresh angles of attack can be identified.

As its 19th-century name suggests—and as many people know from the educational efforts of prominent Parkinson's sufferers such as Janet Reno, Muhammad Ali and Michael J. Fox—the disease is characterized by movement disorders. Tremor in the hands, arms and elsewhere, limb rigidity, slowness of movement, and impaired balance and coordination are among the disease's hallmarks. In addition, some patients have trouble walking, talking, sleeping, urinating and performing sexually.

These impairments result from neurons dying. Although the victim cells are many and found throughout the brain, those producing the neurotransmitter dopamine in a region called the substantia nigra are particularly hard-hit. These dopaminergic nerve cells are key components of the basal ganglia, a complex circuit deep within the brain that fine-tunes and coordinates movement [see box on page 4]. Initially the brain can function normally as it loses dopaminergic neurons in the substantia nigra, even though it cannot replace the dead cells. But when half or more of these specialized cells disappear, the brain can no longer cover for them. The deficit then produces the same effect that losing air traffic control does at a major airport. Delays, false starts, cancellations and, ultimately, chaos pervade as parts of the brain involved in motor control—the thalamus, basal ganglia and cerebral cortex—no longer function as an integrated and orchestrated unit.

Proteins Behaving Badly

IN MANY PARKINSON'S CASES, the damage can be seen in autopsies as clumps of proteins within the substantia nigra's dopaminergic neurons. Such protein masses also feature in Alzheimer's and Huntington's—but in

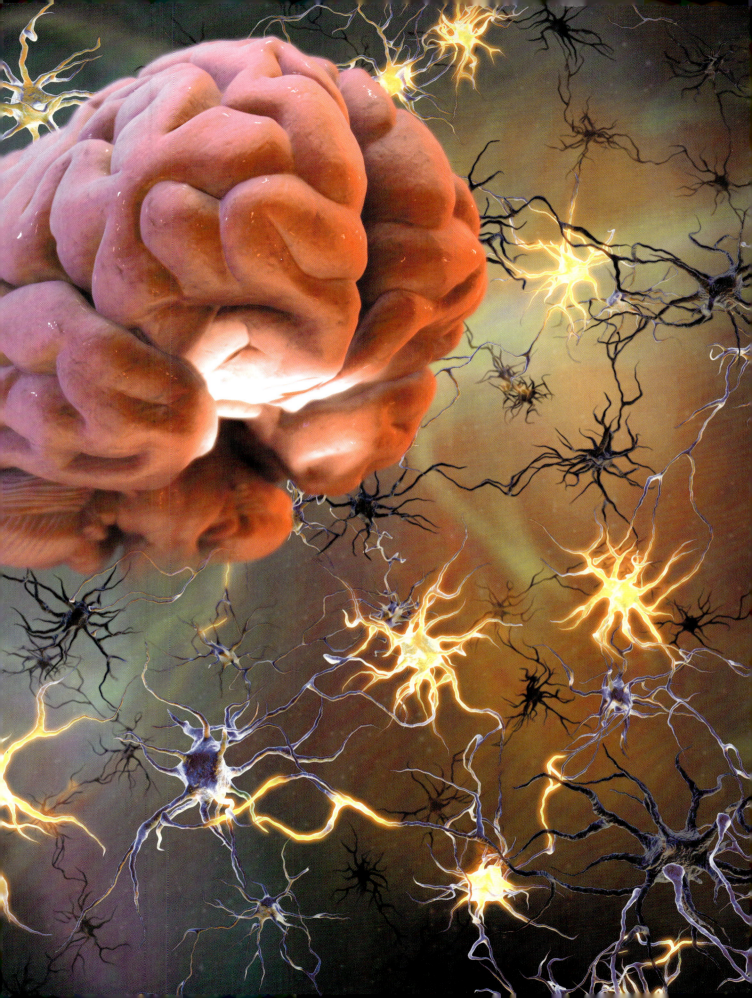

Parkinson's they are called Lewy bodies, after the German pathologist who first observed them in 1912. Like researchers studying those other neurodegenerative diseases, Parkinson's investigators heatedly debate whether the protein clusters themselves cause destruction or are protective and endeavoring to remove toxic molecules from the neurons. Regardless of their position, however, most agree that understanding these accumulations is key to understanding Parkinson's.

Two cellular processes occupy a central place in this emerging story: protein folding and protein elimination. Cells synthesize proteins, which are chains of amino acids, based on instructions written in the DNA of genes. As the proteins are produced, molecules called chaperones fold them into the three-dimensional form they are supposed to take. These chaperones also refold proteins that have become unfolded.

If the chaperone system fails for some reason, proteins not properly folded in the first place or those that did not correctly refold become targeted for disposal by what is called the ubiquitin-proteasome system. First, ubiquitin, a small protein, is attached to a misshapen protein in a process called ubiquitinylation. Such tagging is repeated until ubiquitin chains of varying lengths end up draped over the ill-fated protein. These chains become the kiss of death. They alert the nerve cell's proteasome, a garbage disposal system, to the existence of the bedecked protein. The proteasome then digests it into its constituent amino acids. Aaron Ciechanover and Avram Hershko of Technion-Israel Institute of Technology and Irwin Rose of the University of California at Irvine were awarded the

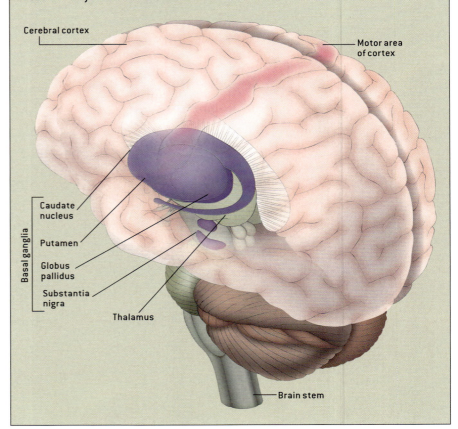

BRAIN REGIONS AFFECTED BY PARKINSON'S

Most cell death occurs in the substantia nigra, which controls voluntary movement and helps to regulate mood. Although the rest of the brain can initially compensate, it can no longer do so when 50 to 80 percent of the cells in the substantia nigra have been lost. At that point, other parts of the brain engaged in motor control, including the rest of the basal ganglia (of which the substantia nigra is part), the thalamus and the cerebral cortex, can no longer work together, and movement becomes disjointed and uncontrollable.

Cerebral cortex

Motor area of cortex

Basal ganglia
Caudate nucleus
Putamen
Globus pallidus
Substantia nigra
Thalamus

Brain stem

2004 Nobel Prize in Chemistry for their work describing this system.

In the past few years, many scientists have come to believe that Parkinson's emerges when the chaperone and ubiquitin-proteasome systems malfunction.

They reason that the disease process might go something like this: some form of injury to neurons of the substantia nigra triggers a cascade of cellular stresses [see "Understanding Parkinson's Disease," by Moussa B. H. Youdim and Peter Riederer; SCIENTIFIC AMERICAN, January 1997]. These stresses result in a wealth of misfolded proteins that congregate. This buildup might initially be protective because all the renegade proteins are herded together and thus prevented from causing trouble elsewhere in a cell. Chaperones then set to work refolding, and the disposal system starts eliminating those proteins that cannot be reformed. When the production of poorly folded proteins overwhelms the cell's

Overview/*Proteins and Parkinson's*

- One of the most pervasive neurological diseases, Parkinson's disease cannot be slowed, stopped or prevented. The two standard forms of treatment, medications and surgery, can only reduce symptoms.
- Recent discoveries about how proteins malfunction and the genetic underpinnings of Parkinson's have opened new avenues for research, and investigators are feeling some optimism about finding new treatments.
- Protein folding and disposal systems gone awry now appear to be central to the disorder, and the genetic causes for those failures have come to light.

ability to process them, however, trouble arises: The ubiquitin-proteasome system becomes inhibited, chaperones get depleted, and toxic proteins accumulate. Neuronal cell death follows.

Researchers espousing this hypothesis think it could explain Parkinson's two forms. An estimated 95 percent of patients suffer from sporadic disease—the results of a complex interplay between genes and the environment. When someone with a susceptible genetic background encounters certain environmental factors, such as pesticides or other chemicals [*see box on this page*], the cells in that individual's substantia nigra suffer more stress and accumulate more misfolded proteins than do the same cells in other people. In the remaining 5 percent of patients, Parkinson's appears to be controlled almost entirely by genetics. Discoveries in the past eight years have revealed a connection between mutations and either the buildup of misshapen proteins or the failure of the cell's protective machinery. These genetic insights have been the most exciting developments in the field in years.

The Genetic Frontier

AT THE NATIONAL Institutes of Health in 1997, Mihael H. Polymeropoulos and his colleagues identified a mutation in a gene for a protein called alpha-synuclein in Italian and Greek families with an inherited form of Parkinson's. It is an autosomal dominant mutation, meaning just one copy (from the mother or the father) can trigger the disease. Mutations in the *alpha-synuclein* gene are extremely rare and insignificant in the worldwide burden of Parkinson's (they account for far less than 1 percent of patients), but identification of the link between the encoded protein and Parkinson's set off an explosion of activity—in part because alpha-synuclein, normal or otherwise, was soon found to be one of the proteins that accumulates in the protein clumps. Investigators reasoned that a better understanding of how the mutation leads to Parkinson's could suggest clues to the mechanism underlying Lewy body formation in dopamine-producing cells of the substantia

nigra in patients with sporadic disease.

The *alpha-synuclein* gene codes for a very small protein, only 144 amino acids long, which is thought to play a role in signaling between neurons. Mutations result in tiny changes in the amino acid sequence of the protein—in fact, several such mutations are now known, and two of them result in the change of a single amino acid in the sequence. Studies of fruit flies, nematodes (roundworms) and mice have shown that if mutated alpha-synuclein is produced in high amounts, it causes the degeneration of dopaminergic neurons and motor deficits. Other studies have revealed that mutated alpha-synuclein does not fold correctly and accumulates within Lewy bodies. Altered alpha-synuclein also inhibits the ubiquitin-proteasome system and resists proteasome degradation. In addition, it has recently become clear that having extra copies of the normal *alpha-synuclein* gene can cause Parkinson's.

In 1998, one year after the discovery of the *alpha-synuclein* mutation, Yoshikuni Mizuno of Juntendo University and Nobuyoshi Shimizu of Keio University, both in Japan, identified a second gene, *parkin*, that is mutated in another familial form of Parkinson's. This mutation appears most often in individuals diagnosed before age 40; the younger the age of onset, the more likely the disease is caused by a *parkin* mutation. Although people who inherit a defective copy from both parents (that is, when the mutation is autosomal recessive) inevitably develop the disease, those who carry a single copy of the mutated gene are also at

---

ENVIRONMENTAL CULPRITS

The idea that Parkinson's disease may be caused by something in the environment has been around for decades. But proof came only in the early 1980s, when J. William Langston of the Parkinson's Institute in Sunnyvale, Calif., studied a group of drug abusers in the San Francisco Bay Area. These young addicts had developed parkinsonian symptoms within days of taking China white, a synthetic heroin. It turned out that the batch contained an impurity called MPTP, a compound that can kill neurons in the brain's substantia nigra region. Through treatment, some of the "frozen addicts," as they came to be called, recovered some movement control; in most, however, the effects were irreversible.

In subsequent years, investigators searched for other compounds with similar effects, and in 2003 their work was bolstered when the National Institute for Environmental Health Sciences put $20 million behind efforts to identify and study environmental causes of Parkinson's. To date, epidemiological and animal studies have linked some cases to high exposure to various pesticides, herbicides and fungicides, including paraquat and maneb. J. Timothy Greenamyre of Emory University has also discovered in animal studies that exposure to rotenone,

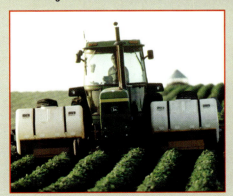

a pesticide often used in organic farming because it is made from natural products, is capable of inducing protein aggregation, killing dopamine-producing neurons, inhibiting the cells' energy-producing organelles and giving rise to motor deficits.

Just as some agents may trigger Parkinson's, others might confer protection. Experts now accept that smoking and coffee drinking can be somewhat protective—although clearly the risks of smoking far outweigh this particular benefit.

—*A.M.L. and S.K.K.*

SOME PESTICIDES, including one routinely used in organic farming, can induce parkinsonian conditions in animals.

CURRENT THERAPIES

Physicians take two basic approaches to treating Parkinson's disease. Both can produce striking benefits, but they also have disadvantages, which is why patients and researchers are so eager for new strategies.

MEDICATIONS

The principal treatments encompass medications that mimic dopamine, compounds used to create dopamine in the brain (such as levodopa), and drugs that inhibit the breakdown of dopamine. Several others act on some of the nondopamine systems affected in Parkinson's, including those mediated by the neurotransmitters acetylcholine and glutamate. Many of these drugs help during the initial phases of the disease, but their ongoing use can become problematic. Chief among the long-term adverse effects are unpredictable oscillations between periods of good motor function and periods of freezing, tremor and rigidity. In addition, some of the medications can cause involuntary twisting, writhing movements called dyskinesias, which are particularly prominent in young patients and are extremely disabling.

DEEP-BRAIN STIMULATION

At the turn of the century, investigators discovered that destroying a small number of cells in the brain's motor pathways could reduce parkinsonian tremors. Although the procedure often caused muscle weakness, patients preferred that to the shaking. Then, in 1938, surgeons injured the basal ganglia and found even more marked improvement in Parkinson's patients. It appeared that eliminating the cells that were misbehaving—that is, those misfiring or firing too much—apparently allowed the rest of the brain to function normally. Unfortunately, creating lesions was not a perfect solution. If they were not precisely placed or if they involved both sides of the brain, they could cause severe damage, impairing speech and leading to cognitive problems.

In the 1970s investigators discovered that high-frequency electrical stimulation of parts of the brain could mimic lesions, without reproducing the side effects. Various forms of deep-brain stimulation are used for many neurological disorders today [see "Stimulating the Brain," by Mark S. George; SCIENTIFIC AMERICAN, September 2003]. In Parkinson's patients, an electrode is placed in one of two basal ganglia targets—the globus pallidus or subthalamic nucleus—and attached to a pulse-generating device implanted in the chest (below). The pacemaker typically delivers 90-microsecond, three-volt pulses of electricity up to 185 times per second and needs to be replaced every five years.

The pioneers of the technique, Alim Louis Benabid and Pierre Pollak of the University of Grenoble in France, report that such stimulation dramatically reduces tremor and rigidity. Indeed, in the past decade or so it has become a mainstay of treatment, and an estimated 30,000 patients have undergone the surgery. Some have been able to reduce the doses of medicines they take, whereas others have stopped taking them altogether. At the same time, however, deep-brain stimulation cannot prevent the disease from progressing, and it cannot alleviate the problems with cognition, speech and balance that may arise.

Despite the success of deep-brain stimulation, many questions remain. For one thing, it is not clear whether the globus pallidus or the subthalamic nucleus is a better target. In addition, the precise electrical and chemical mechanisms by which electrical energy improves Parkinson's disease remain to be determined, and much of the data are conflicting. For example, researchers used to think deep-brain stimulation worked the same way lesioning did, by inactivating cells. Recently, however, they have learned that the procedure seems to cause faster firing of impulses.   —A.M.L. and S.K.K.

Electrode

Basal ganglia

Implanted wire

Implanted pacemaker

---

greater risk. *Parkin* mutations appear to be more common than *alpha-synuclein* gene mutations, but no good figure on incidence is currently available.

The parkin protein contains a number of amino acid sequences, or domains, common to many proteins. Of particular interest are two so-called RING domains; proteins with these RING domains are involved in the protein degradation pathway. Findings now suggest that neuronal death in this form of Parkinson's stems in part from the failure of the ubiquitinylation component of the protein disposal system: parkin attaches ubiquitin to misfolded proteins—without it, there is no tagging and no disposal. Our own work has recently shown that a protein called BAG5, which is found in Lewy bodies, can bind to parkin to inhibit its function and cause the death of dopamine-producing neurons.

Interestingly, some patients with *parkin* mutations lack Lewy bodies in their nigral neurons. This observation suggests that proteins may not form aggregates unless the ubiquitinylation process is functioning. It also suggests that when harmful proteins are not huddled together within Lewy bodies they create cellular havoc. Because patients with *parkin* mutations develop the disease early in life, it seems likely that they miss some initial protection conferred by having toxic proteins quarantined in clumps.

Several other recent discoveries highlight further genetically induced muck-ups in the cellular machinery. In 2002 Vincenzo Bonifati and his colleagues at Erasmus Medical Center in Rotterdam identified a mutation in a gene called *DJ-1*. Like that in *parkin*, this mutation is responsible for an autosomal recessive form of Parkinson's and has been found in Dutch and Italian families. Investigators have seen mutations in another gene, *UCHL1*, in patients with familial Parkinson's. A paper in *Science* just described a mutation in *PINK1* that may lead to metabolic failure and cell death in the substantia nigra. And other work has identified a gene called *LRRK2*, which encodes the protein dardarin (meaning "tremor" in the Basque region, where the affected patients came from). It, too, is involved with metabolism and appears in familial Parkinson's. But researchers are not far along in understanding exactly what all these mutations set wrong.

New Avenues for Treatment

BECAUSE THE INSIGHTS just described involve molecules whose activity could potentially be altered or mimicked by drugs in ways that would limit cell death, the discoveries could lead to therapies that would do more than ease symptoms—they would actually limit the neuronal degeneration responsible for disease progression.

This strategy has yielded two intriguing results. Increasing the levels of chaperones in cells of the substantia nigra has been found to protect against the neurodegeneration set in motion by mutated alpha-synuclein in animals. Recent studies using fruit-fly models of Parkinson's have shown that drugs that induce chaperone activity can offer protection against neurotoxicity. Perhaps one day chaperone-type drugs can be developed to limit degeneration in people, or gene therapy could be devised to trigger the production of needed chaperones. In addition, investigators have found that increasing the amount of normal parkin protein in cells protects against the neurodegeneration resulting from noxious, misfolded proteins. Much more study will be needed, however, to

determine whether such interventions could be made to work in humans.

In addition to pursuing the preliminary leads that have arisen out of the new protein-related and genetic findings, investigators have begun introducing neurotrophic factors—compounds promoting neuronal growth and differentiation—into the brain. These agents not only alleviate symptoms but also promise to protect neurons from damage or even to restore those already harmed.

One line of research in animals, for instance, suggests that a family of proteins called glial cell line-derived neurotrophic factor (GDNF) can enhance the survival of injured dopamine neurons and dramatically reduce parkinsonian symptoms. Steve Gill and his colleagues at Frenchay Hospital in Bristol, England, have embarked on a pilot study to give Parkinson's patients GDNF. Surgeons insert a catheter into the left and right striatum, the main recipients in the basal ganglia of the dopamine secreted by neurons of the substantia nigra. Minute volumes of GDNF are then continuously infused to the brain from a pump set into the abdomen. The pump holds enough GDNF to last one month and is replenished during an office visit; a syringe pierces the skin and refills the pump reservoir.

Initial results in a handful of patients suggested that symptoms had improved, and PET scans indicated some restoration in dopamine uptake in the striatum and substantia nigra. But the results of a larger, more recent trial have been unconvincing: patients who received saline solution fared no better than those who received GDNF. Nevertheless, many of

us who work in this area feel that this approach is still worth pursuing. It is not unusual in medicine for the first forays into a treatment to be negative. Levodopa, for instance, initially showed no benefit and only unwanted side effects; now it is one of the principal treatments for Parkinson's.

Other researchers are using gene therapy instead of surgery to administer GDNF, hoping the delivered gene will provide a long-term supply of this neurotrophic agent. Jeffrey H. Kordower of Rush Presbyterian-St. Luke's Medical Center in Chicago and Patrick Aebischer of the Neurosciences Institute at the Swiss Federal Institute of Technology and their colleagues engineered a lentivirus to carry the gene for GDNF and deliver it to dopamine-producing striatal cells in four parkinsonian monkeys. The results were impressive: the monkeys' motor problems significantly diminished, and they were unaffected by a subsequent injection of MPTP, a chemical toxic to dopamine neurons of the substantia nigra. The introduced gene induced cells to make the protein for up to six months, after which the experiments were stopped. Based on these studies, scientists at Ceregene in San Diego are using a similar technique to deliver the protein neurturin, a member of the GDNF family. Although the studies are in the preclinical phase, researchers plan to test a gene similar to the gene for neurturin in human patients.

Still other forms of therapy are being investigated. Working with Avigen near San Francisco, Krys Bankiewicz has shown in animals that placing the

Perhaps one day CHAPERONE-TYPE DRUGS can be developed to limit degeneration in people.

THE AUTHORS

ANDRES M. LOZANO and *SUNEIL K. KALIA* have worked together for several years, studying various aspects of Parkinson's disease. Lozano, who was born in Spain and obtained his M.D. from the University of Ottawa, is professor and R. R. Tasker Chair in Stereotactic and Functional Neurosurgery at the Toronto Western Hospital and the University of Toronto. He has devoted his career to understanding the causes of Parkinson's and developing novel surgical treatments. Kalia recently completed his doctoral degree working with Lozano. His research focused on the role of chaperone molecules in Parkinson's.

gene for an enzyme called aromatic amino acid decarboxylase in the striatum can enhance dopamine production in this area of the brain. In rats and monkeys this approach has also ameliorated parkinsonian symptoms. Trials in patients have been approved and will be launched soon.

Michael Kaplitt of Cornell University and his team are taking a different tack, using gene therapy to shut down some of the brain regions that become overactive when dopamine released from the substantia nigra falls too low—including the subthalamic nucleus of the basal ganglia. (The loss of dopamine causes neurons making glutamate, an excitatory neurotransmitter, to act unopposed and thus overstimulate their targets, causing movement disorders.) Kaplitt will begin human trials using a virus to introduce the gene for glutamic acid decarboxylase— which is crucial to the production of the inhibitory neurotransmitter gamma amino butyric acid (GABA)—to these sites. He and his co-workers hope that the GABA will quell the overexcited cells and thus calm parkinsonian movement disorders. In the experiments, they thread a tube about the width of a hair through a hole the size of a quarter on top of a patient's skull. The tube delivers a dose of virus, which ferries copies of the gene into neurons of the subthalamic nucleus. The chemical released from the altered cells should not only quiet the overactive neurons residing in that region but may be dispatched to other overactive brain areas.

Perhaps the most hotly debated potential treatment entails transplanting cells to replace those that have died. The idea has been to implant embryonic stem cells or adult stem cells and to coax these undifferentiated cells into becoming dopamine-producing neurons. Because embryonic stem cells are derived from days-old embryos created during in vitro fertilization, their use is highly controversial. Fewer ethical questions surround the use of adult stem cells, which are harvested from adult tissue, but some scientists believe these cells are more difficult to work with.

Despite important progress in iden-

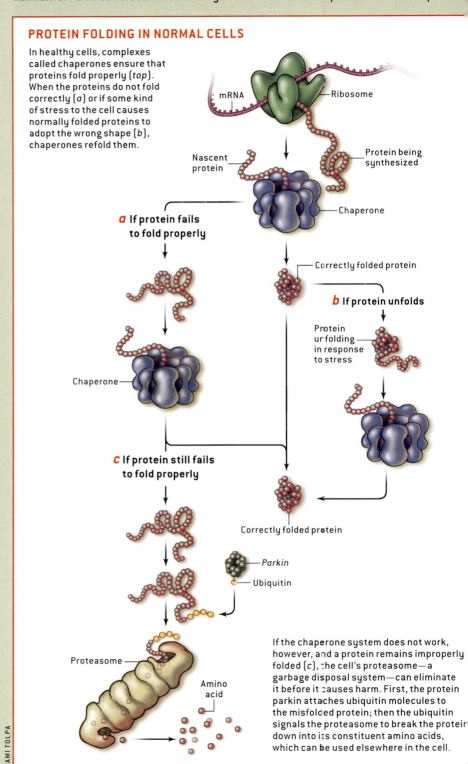

PROTEINS AND PARKINSON'S

Accumulations of misfolded proteins called Lewy bodies have been recognized for decades as a hallmark of Parkinson's. Scientists do not yet know whether these protein clusters are protecti

PROTEIN FOLDING IN NORMAL CELLS

In healthy cells, complexes called chaperones ensure that proteins fold properly (top). When the proteins do not fold correctly (a) or if some kind of stress to the cell causes normally folded proteins to adopt the wrong shape (b), chaperones refold them.

mRNA
Ribosome
Nascent protein
Protein being synthesized
Chaperone

a If protein fails to fold properly

Correctly folded protein

b If protein unfolds

Protein unfolding in response to stress

Chaperone

c If protein still fails to fold properly

Correctly folded protein

Parkin
Ubiquitin

Proteasome

Amino acid

If the chaperone system does not work, however, and a protein remains improperly folded (c), the cell's proteasome—a garbage disposal system—can eliminate it before it causes harm. First, the protein parkin attaches ubiquitin molecules to the misfolded protein; then the ubiquitin signals the proteasome to break the protein down into its constituent amino acids, which can be used elsewhere in the cell.

TAMI TOLPA

cause they keep the toxic proteins out of mischief) or whether they ultimately trigger the death
erve cells. Nevertheless, it is clear that proteins gone awry underlie this devastating disease.

WHAT GOES WRONG IN PARKINSON'S

For reasons not fully understood, the chaperone and proteasome system fail in people who
become ill with Parkinson's. Misfolded proteins accumulate in cells because the chaperones
cannot keep up or the proteasome system cannot break down the miscreant proteins fast
enough; this buildup can damage and kill affected neurons. Recent genetic studies have
suggested that mutant forms of two proteins—alpha-synuclein (*left*) and parkin (*right*)—can
help undermine the chaperone and protein disposal system.

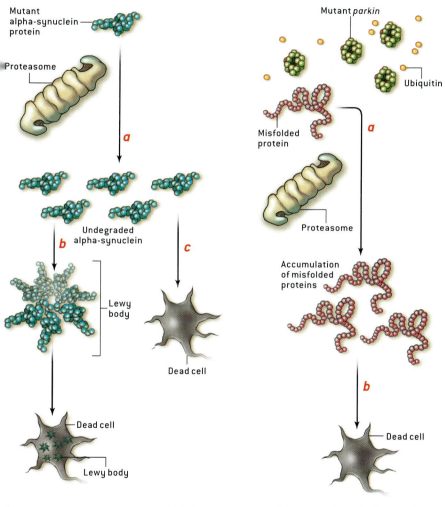

A very rare mutation in the *alpha-synuclein*
gene can cause Parkinson's by giving
rise to a form of the protein that resists
breakdown by proteasomes (*a, above*). In
a sign that Lewy bodies might sometimes
be protective, groups of mutant alpha-
synuclein that end up in a Lewy body (*b*)
appear to be less damaging initially than
copies of the protein that roam the nerve
cell, causing its quick demise (*c*).

In the case of parkin, the mutated
versions fail to add ubiquitin to
misfolded proteins. As a result, the
proteasome cannot break down the
proteins (*a, above*), which ultimately
cause cell death (*b*). Mutated parkin
does not give rise to Lewy bodies.

tifying the molecular cues and recipes
for pushing undifferentiated cells to produce dopamine, no one yet knows
whether transplantation of any kind will
be as fruitful a strategy as has been
hoped. The clinical trials using the most
meaningful protocols have so far been
conducted with fetal material. These
have shown hundreds of thousands of
surviving transplanted dopamine-producing cells in patients, yet the functional benefits have been at best modest and
inconsistent, and the treatment has been
associated with serious adverse effects,
including dyskinesias (uncontrollable
writhing and twisting movements). Scientists are trying to determine why
transplantation has not been more helpful and why side effects have arisen, but
for now they are not conducting human
trials of the procedure in the U.S.

Finally, researchers continue to investigate and refine the approach behind
deep-brain stimulation: applying elec-
tric pulses. Several months ago Stéphane
Palfi and his colleagues at the CEA Fré-
déric Joliot Hospital Service in Orsay,
France, reported that gently stimulating
the brain surface could improve symptoms in baboons with a version of Parkinson's. Clinical trials are under way in
France and elsewhere to determine
whether this surgical intervention is similarly effective in humans.

Although much remains unknown
about Parkinson's, the genetic and cellular insights that have come to light in
just the past few years are highly encour-
aging. They give new hope for treatments
that will combine with existing ones to
slow disease progression and improve
control of this distressing disorder. **SA**

MORE TO EXPLORE

Parkinson's Disease, Parts 1 and 2. A. E.
Lang and A. M. Lozano in *New England Journal
of Medicine,* Vol. 339, pages 1044–1053 and
pages 1130–1143; October 8 and October
15, 1998.

**Genetic Clues to the Pathogenesis of
Parkinson's Disease.** Miguel Vila and Serge
Przedborski in *Nature Medicine,* Vol. 10,
pages S58–S62; July 2004.

**Neurodegenerative Diseases: A Decade of
Discoveries Paves the Way for Therapeutic
Breakthroughs.** Mark S. Forman, John Q.
Trojanowski and Virginia M-Y. Lee in *Nature
Medicine,* Vol. 10, pages 1055–1063; 2004.

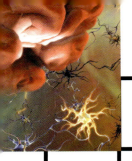

Article Review: CELL BIOLOGY

New Movement in Parkinson's   *by Andres M. Lozano and Suneil K. Kalia*

TESTING YOUR COMPREHENSION

1. In Parkinson's disease, dopamine producing cells are lost as the disease progresses. Define the location of these cells, and the role these cells play in normal brain function.

   ANSWER: *Dopaminergic neurons are located in the substantia nigra, a key structure within the basal ganglia neuronal network. The basal ganglia plays a key role in the coordination and organization of signals that regulate movement. The loss of these cells results in the "shaking palsy" phenotype.*

2. What is the current estimated number of Parkinson's cases in the world?

   ANSWER: *4 million*

3. The proteolytic destruction of improperly folded proteins is a normal cellular function, destroying the "cellular trash". Describe the process of identifying a protein as improperly folded, tagging that protein, and the mechanism of destruction.

   ANSWER: *Molecular chaperones mediate protein folding, identifying misfolded proteins and facilitating their proper folding. When a protein cannot be properly folded, the chaperone works with ubiquitin–proteasome system (UPS). The misfolded protein is conjugated to ubiquitin, which facilitates recognition of that protein by the proteasome. The proteasome enzymatically cuts the protein into peptide fragments.*

4. The addition of ubiquitin to misfolded proteins results in their degradation. Is ubiquitin a sugar, a lipid, or a protein? What enzyme(s) mediate the addition of ubiquitin to misfolded proteins?

   ANSWER: *Ubiquitin is a small protein (76 amino acids), that is added to misfolded proteins. This article mentions the essential role of the parkin protein in ubiquitinylation of misfolded proteins. There are additional enzymes that are involved in this process including ubiquitin activating enzyme (E1), ubiquitin–conjugation enzyme (E2), and ubiquitin ligase (E2-E3).*

CLASS ACTIVITIES AND DISCUSSION

1. Describe the progression of physiogical stress in a neuron that completely lacks the ability to ubiquitinate and destroy misfolded proteins. A helpful analogy would be a manufacturing center that has an inability to destroy/recycle non-functional products. Include chaperone proteins as a component of your discussion.

2. The endoplasmic reticulum (ER) is a key site of cellular protein synthesis. Cellular signals emerge from the ER to activate the "unfolded protein response". Examine this signaling mechanism using Entrez PubMed at the National Library of Medicine (www.pubmed.org). You will rapidly identify many articles on this topic, try to refine your search by using different keywords and focusing your attention to a specific component of the UPR (apoptosis, kinase signaling, calcium, chaperones, etc.).

3. The cell death that is activated during Parkinson's disease is known as programmed cell death or apoptosis. How is this cell death different from necrotic cell death?

4. It is thought that in many cases, an initial insult to the brain begins the progression towards Parkinson's in susceptible individuals. What types of insults are known, and what types of insults are hypothesized to contribute to the initiation of some types of Parkinson's disease? Is Alzheimer's disease triggered by a physical insult to the brain?

5. If you worked for a federal agency that had oversight over pesticide exposure, how might you develop guidelines for establishing "safe" levels of different pesticides? What types of laboratory tests would you feel were appropriate predictors of neurotoxicity?

Article Review: GENETICS

TESTING YOUR COMPREHENSION

1. Define the relevance of the α-synuclein gene/protein in Parkinson's disease.

   ANSWER: *The α-synuclein protein is associated with the protein clumps found within the neurons of Parkinson's patients. Mutations in the α-synuclein gene are associated with an inherited predisposition to Parkinson's disease (autosomal dominant). In cellular studies, the overexpression of mutant α-synuclein is sufficient to cause the death of dopaminergic neurons.*

2. Define the relevance of the parkin gene/protein in Parkinson's disease.

   ANSWER: *Mutations in the parkin gene are associated with individuals that suffer from Parkinson's disease. This trait is inherited as an autosomal recessive though carriers have an increased tendency to develop the disease. The parkin protein appears to play a central role in ubiquitinylation of misfolded proteins.*

3. One valuable tool in the characterization of Huntington's disease is a toxin that induces cell death within the substantia nigra region of the brain. Name this toxin and how its impact upon the brain was initially discovered.

   ANSWER: *MPTP is a potent toxin that induces cell death in dopaminergic neurons. This effect upon neurons was discovered in the 1980s. Individuals prepared a "designer drug" – synthetic heroin, MPPP – for recreational use, but the final product contained trace amounts of MPTP as a by product of the synthesis reaction. People who used this drug displayed a rapid progression into a Parkinson's-like disease, and scientists found severe damage to the substantia nigra of these patients.*

CLASS ACTIVITIES AND DISCUSSION

1. With the recent discoveries of some of the key genes involved in Parkinson's disease, why is a cure still anticipated to be some time away? Discuss the progression from recognizing the cellular process behind a disease to developing effective therapies. Do you feel the public recognizes the challenges to developing these types of therapies?

2. Break the class into 6 groups, and assign each group one of the Parkinson's therapies mentioned in this article. Each group will research the fundamental technologies of their assigned therapy, the theorized or realized benefit of this therapy to Parkinson's sufferers, and the limitations of this therapy. Does the therapy offer a cure, or only a relief from Parkinson's symptoms? Does the therapy affect the cellular mechanisms that cause Parkinson's (i.e. protein clumps and cell death) or the symptoms of the disease? Suggested topics for each group are 1) drugs that induce elevated chaperone expression, 2) catheter-based delivery of neurotrophic factors (GDNF), 3) gene-therapy based delivery of neurotrophic factors, 4) gene-therapy based inhibition of brain regions that cause tremors, 5) use of stem cells to replace dopamine-producing cells, and 6) administration of drugs that mimic dopamine or elevate dopamine production within the brain.

3. The Michael J. Fox Foundation for Parkinson's Research was founded and organized to support research into Parkinson's disease. Review their website (www.michaeljfox.org) and examine the funding initiatives described there. What key areas of Parkinson's research are funded through this foundation? In comparison to the National Institutes of Health, this foundation funds a rather small collection of researchers. Can the efforts of an individual offer a significant benefit to the study of this complex disease? Provide evidence for your answer. Can you find another example of a small, but active research foundation working to support medical research?

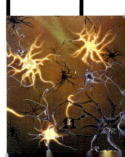

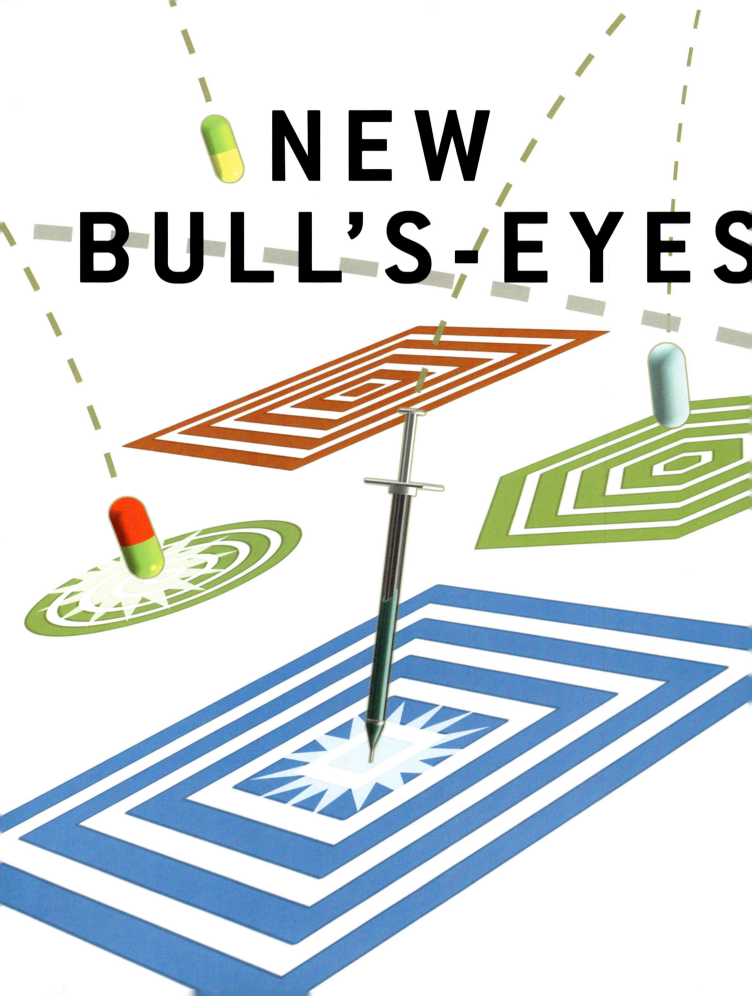

NEW BULL'S-EYES

A familiar class of cell-surface receptors turns out to offer an array of fresh targets that could yield new treatments for disorders ranging from HIV infection to obesity

FOR DRUGS

By Terry Kenakin

An amazing fraction—roughly half—of all the medicines prescribed today have a striking commonality. At the molecular level, they act on the same type of target: a serpentine protein that weaves seven times through the membrane that envelops the cell. External parts of each serpent serve as an antenna for molecular signals approaching the cell, and internal parts trigger the cell's responses to such cues, beginning with the activation of a signal processor called a G-protein. The serpents themselves are thus known as G-protein coupled receptors, or GPCRs.

As a group, GPCRs show far more versatility than any other class of cell-surface receptor. For instance, the natural molecules to which GPCRs respond range in size from neurotransmitters that are only a few times as massive as a single carbon atom all the way up to proteins 75 times larger than that. Moreover, GPCRs participate in just about every bodily function that sustains life, from heartbeat and digestion to breathing and brain activity. The drugs that target these receptors are equally diverse. The list includes blood pressure reducers (such as propranolol), stomach acid suppressors (such as ranitidine), bronchodi-

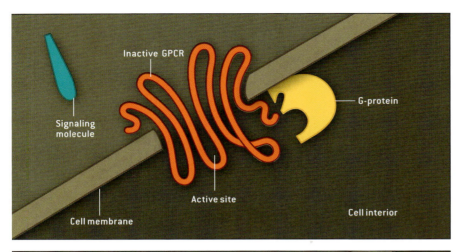

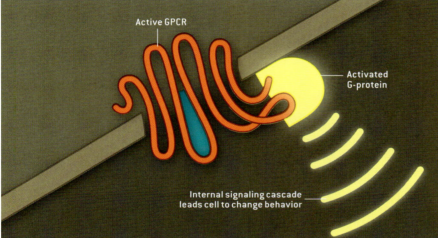

G-PROTEIN COUPLED RECEPTOR (GPCR), which snakes through the cell membrane seven times, typically issues no messages into a cell (*top*) until a signaling molecule, such as a hormone or a neurotransmitter, binds to a region called the active site. This binding (*bottom*) causes the receptor to activate a molecule called a G-protein, which triggers a series of intracellular interactions culminating in a change in the cell's behavior. New insights into the functioning of GPCRs suggest novel avenues for treating disease.

words, despite the wealth of medicines already known to act on these fascinating receptors, many more may lie ahead. The search for such pharmaceuticals is still in early stages, but a few agents, including some for HIV infection (the cause of AIDS), are now advancing through human trials.

Shape Matters

UNTIL ABOUT 10 YEARS AGO, pharmaceutical researchers thought that to influence the activity of GPCRs they would have to aim drugs at a receptor's active site. During the body's normal operation, a neurotransmitter or other information-bearing molecule (or "ligand") at the cell's outer surface essentially plays the "key" to the active site's "lock." So a substance that plugged the lock could prevent unwanted signaling through the receptor by any key and serve as an inhibitor. Conversely, something that mimicked the natural ligand could essentially open the lock and therefore take the place of the natural key if it were missing.

Scientists thought as well that the best way to evoke a selected physiological response was to choose a compound that interacted with a specific form of a receptor but ignored other variants. The neurotransmitter norepinephrine, for instance, activates two types of GPCR, called alpha and beta adrenoceptors, of which the first has four subtypes and the second has three. These various receptors, in turn, govern different life-sustaining processes. In the heart, beta$_1$ adrenoceptors quicken the heart rate and increase the force of each beat; in the lungs, beta$_2$ adrenoceptors widen the air passageways. Hence, to open constricted airways without unwanted

lators (such as albuterol) and antidepressants (such as paroxetine). The disorders these medicines treat include hypertension, congestive heart failure, ulcer, asthma, anxiety, allergy, cancer, migraine and Parkinson's disease.

Incredibly, today's GPCR-targeting drugs all work in one of two ways—they either attach to the "antenna" region of the receptor (also known as the active site) and mimic the effect of the natural neurotransmitter, hormone or other molecule that normally signals through the GPCR, or they interfere with a natural signaler's ability to act on the antenna. Over the past 15 years, a technological revolution has furnished investigators with new eyes with which to see GPCRs at work. Consequently, other ways of manipulating GPCR activity have emerged and are beginning to be mined for drug discovery. In other

Overview/*New Drug Targets*

- Proteins called G-protein coupled receptors (GPCRs), which sit on the cell surface, convey signals from hormones and the like into cells by activating G-proteins—signal processors residing just under the cell membrane.
- About half of all pharmaceuticals on the market act on GPCRs, binding to the sites normally targeted by the body's own extracellular signalers.
- In the past 10 years, researchers have learned that GPCR activity can also be modulated by compounds that bind to other sites on GPCRs. This discovery opens new possibilities for treating cancer and other major disorders.

effects on the heart, pharmaceutical makers might seek an agent that mimicked norephinephrine's ability to stimulate beta$_2$ adrenoceptors but without binding to beta$_1$ adrenoceptors.

Many medicines do, in fact, function as inhibitors or agonists (mimics) by interacting with the active site of a specific GPCR. But an emerging drug development strategy has to do with the "allosteric" nature of GPCRs: the shape of one part of the receptor can affect the conformation, and thus the activity, of a distant part.

GPCRs constantly adopt somewhat different shapes, essentially sampling a library of conformations. When a natural signaling molecule binds to the active site, it stabilizes the arrangement that activates G-proteins. But it turns out that certain molecules, known as allosteric modulators, can bind elsewhere to influence form and activity. Some stabilize GPCR conformations that promote signaling, whereas others maintain shapes that impede it (say, by burying the active site so that it becomes inaccessible to its natural ligand).

The implications are profound. The entire receptor can theoretically offer binding sites, at any one of which a diminutive molecule might stabilize a shape that yields some biological effect. This property greatly enlarges the vista for therapeutic modification of GPCR function.

AIDS researchers are among those actively pursuing the potential of allosteric modulators, trying to find ones able to block HIV from infecting cells. Biologists have long known that the virus attacks cells called helper T lymphocytes by adhering to a cell-surface protein named CD4. But in the mid-1990s they learned that this protein does not act alone.

To enter cells, the virus also has to bind to an additional anchor: a GPCR known as CCR5 (or, in late-stage infection, a GPCR called CXCR4). Normally CCR5 responds to any of three chemokines, natural signals that can attract immune system cells to a site of infection. Unfortunately, it also offers a hook for the virus's coat protein (gp120). In-

deed, CCR5 now appears to be a central player in HIV infection; people whose genetic makeup causes them to lack a functional form tend to be extraordinarily resistant to HIV.

Several allosteric modulators that hold CCR5 in a shape inimical to binding by HIV's gp120 have already reached human trials. Blocking the gp120-CCR5 interaction by delivering these tiny drugs is an achievement comparable to, in a geophysical analogy, an island the size of Fiji preventing two Australias from coming together. In more allegorical terms, if such drugs work they will be the David that smites Goliath.

Beyond Volume Control

THE EFFECTS produced by GPCRs depend not only on the extracellular molecules that bind to them but also on how many copies of the receptors are accessible on the cell surface. As might be expected, when extracellular signalers bind to many copies of a receptor, the cell receives a "louder" message and undergoes a more pronounced behavioral change than when few copies of

the receptor are bound. But the number of receptors can do more than control "volume." It can actually influence which of several G-protein species become stimulated and can thereby lead to activation of distinct pathways (cascades of molecular interactions) inside a cell.

G-proteins come in four major forms, with subtypes in each class. Each has a different proclivity for working with any given GPCR, and for its part a GPCR may not be equally active toward all G-proteins. A scant supply of a given receptor might therefore result in activation of only the most sensitive G-protein, whereas a greater abundance might lead to responses by multiple G-proteins, eliciting a different cellular behavior.

Accordingly, a GPCR can no longer be seen as simply a toggle switch turned on by a hormone or neurotransmitter and turned off when the natural signal diffuses away from its binding site. It is a much more sophisticated information-processing unit.

Theoretically, the variety of response patterns a given GPCR can generate will

MARKETED DRUGS ACTING ON GPCRs

The products listed below are just a sampling of marketed compounds targeting GPCRs; they act on various receptors.

BRAND NAME (GENERIC NAME) AND MAKER	EFFECT
Allegra (fexofenadine) *Aventis*	Blocks histamine action, to control allergic responses
Duragesic (fentanyl) *Janssen*	Relieves pain
Flomax (tamsulosin) *Boehringer Ingelheim*	Eases symptoms of enlarged prostate
Imitrex (sumatriptan) *GlaxoSmithKline*	Eases migraines
Lopressor (metoprolol) *Novartis*	Lowers blood pressure
Oxycontin (oxycodone) *Purdue*	Relieves pain
Pepcid (famotidine) *Merck*	Counteracts stomach acid
Phenergan (promethazine) *Wyeth*	Blocks histamine
Serevent (salmeterol) *GlaxoSmithKline*	Opens airways
Singulair (montelukast) *Merck*	Controls airway inflammation
Sudafed (pseudoephedrine) *Pfizer*	Eases nasal congestion
Zantac (ranitidine) *GlaxoSmithKline*	Counteracts stomach acid
Zyrtec (cetirizine) *Pfizer*	Blocks histamine
Zyprexa (olanzapine) *Eli Lilly*	Eases symptoms of various psychoses

MANY AVENUES OF ATTACK

Most drugs on the market target the active site of some cell-surface receptor, and many aim for the active site of a specific GPCR (*below*). Yet molecules acting at regions outside the active site can also influence GPCR activity (*right*). Recent studies encourage hope that small molecules targeted to those additional sites could be administered to activate or quiet GPCRs involved in various diseases.

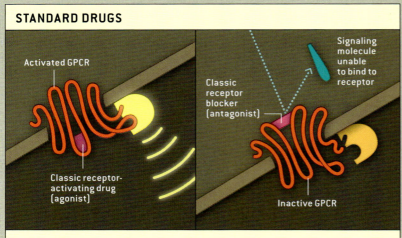

STANDARD DRUGS

Activated GPCR

Classic receptor-activating drug (agonist)

Signaling molecule unable to bind to receptor

Classic receptor blocker (antagonist)

Inactive GPCR

Pharmaceuticals that act on GPCRs today usually fit into the active site and either mimic the action of a natural signaling molecule (*left*) or prevent the native signaler from docking with the receptor and thus from acting on a cell (*right*).

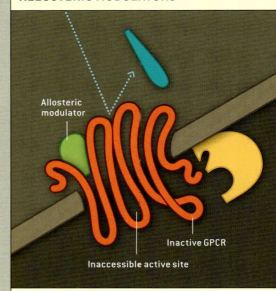

ALLOSTERIC MODULATORS

Allosteric modulator

Inactive GPCR

Inaccessible active site

These agents stabilize the conformation of a GPCR in a way that enhances (*not shown*) or decreases (*above*) the receptor's activity, such as by causing the active site to become inaccessible to a signaling molecule.

depend on both the range of ligands it can detect and the mix of G-protein species it can activate. If, for example, a receptor can detect any of three different signals and can activate any one, two, three or all four of the major G-proteins (as is known to be the case for the GPCR responsive to thyrotropin, the pituitary hormone that stimulates the thyroid gland), the receptor gains the theoretical capacity for dozens of forms of behavior, each seen at one time or another. If it were only a toggle switch, it could have only two.

Research also suggests that drugs can take advantage of this complexity in receptor function. Distinct substances might cause a receptor to hold different biologically active shapes, each of which might interact with a distinct G-protein or G-protein combination, triggering the activity of divergent intracellular paths. Agents that can cause cells to increase or decrease the quantity of receptors at the surface, rather than altering

GPCR activity per se, should be valuable as well.

This last strategy could be pursued for combating HIV. One problem that might arise from relying on allosteric modulators to prevent the viral coat protein from finding its docking site on CCR5 is that the virus mutates rapidly. This mutability could lead to the creation of a coat protein that would bind quite well to an allosterically altered CCR5. A plausible way to avert this threat would be to banish the receptor from the cell surface, thereby denying the virus its point of attack.

Like all other GPCRs, CCR5 is synthesized endlessly by the cell, stationed at the surface and then drawn back inside for degradation or recycling. And certain chemokines are known to promote CCR5 internalization. This observation raises the possibility of finding pharmacological agents that would not only accelerate the removal of CCR5 from the cell surface but would

also serve as therapies to which the virus could not adapt. After all, no change that HIV could undergo would enable it to latch onto CCR5 if that receptor were removed from the cell surface.

Stopping Renegade Signaling

BEYOND BEING controllable by allosteric modulators, GPCRs may exhibit another biologically important behavior, known as constitutive activity—that is, sometimes they activate G-proteins even without being "told" to do so by a bound ligand. As is true in other forms of GPCR functioning, this one arises from a particular shape in the receptor's repertoire. The conformation, however, is one that the receptor rarely takes. Under normal circumstances, the number of molecules that adopt it will therefore be quite small, and so they will have little effect on the cell's overall behavior and will be hard to detect. But if the constitutively active

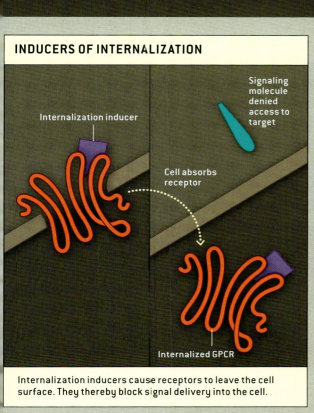

INDUCERS OF INTERNALIZATION

Internalization inducer

Signaling molecule denied access to target

Cell absorbs receptor

Internalized GPCR

Internalization inducers cause receptors to leave the cell surface. They thereby block signal delivery into the cell.

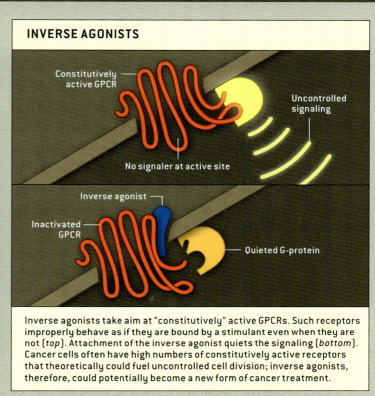

INVERSE AGONISTS

Constitutively active GPCR

Uncontrolled signaling

No signaler at active site

Inverse agonist

Inactivated GPCR

Quieted G-protein

Inverse agonists take aim at "constitutively" active GPCRs. Such receptors improperly behave as if they are bound by a stimulant even when they are not (*top*). Attachment of the inverse agonist quiets the signaling (*bottom*). Cancer cells often have high numbers of constitutively active receptors that theoretically could fuel uncontrolled cell division; inverse agonists, therefore, could potentially become a new form of cancer treatment.

receptors become sufficiently abundant, their combined signaling can exert a powerful influence.

The consequences become especially dramatic in illnesses such as viral infection or cancer, which may advance by inducing one or another receptor to behave in ways that promote the disease. In a form of pancreatic cancer, for example, the receptor for a hormone called vasoactive intestinal peptide (VIP) might be such a bad actor.

In a normal pancreatic cell that displays this GPCR, activation of the receptor by VIP supports cell division. But in people afflicted by this malignancy, the receptor becomes overabundant and the versions that act independently, without need for VIP stimulation, become correspondingly numerous— together they acquire the capacity for driving unconstrained proliferation of tumor cells. Oncologists have long been familiar with destructive constitutive activity in certain non-GPCR recep-

tors, notably one called ras. In those cases, though, mutations in the receptor, rather than an aberrant plentitude of receptors, account for the behavior.

Standard pharmaceuticals cannot quell the cellular misbehavior triggered by constitutively active receptors. A conventional receptor stimulant, or agonist, would only cause more receptors to hold an active shape, to the patient's detriment. A conventional receptor blocker, or antagonist, might prevent natural signals from activating receptors, but such agents will have no effect on receptors that need no outside prompting in order to act. Thus, a new kind of drug is required, one that forces constitutively active GPCRs to maintain an inactive shape.

Such agents, called inverse agonists, might one day constitute an important new form of cancer therapy. They are also being eyed for treating obesity. In this realm, the envisioned targets include the receptor for ghrelin, a recently

discovered hormone produced chiefly by the stomach, and the H_3 subtype of histamine receptor; both receptors appear to participate in the brain's regulation of appetite.

Exploiting Phantom Genes

AT LEAST ONE OTHER FORM of GPCR behavior remains to be mined for drug discovery. Cells sometimes mix and match proteins, forming complexes that function as receptors having sensitivities not seen in the individual components. In the most extreme form of this activity, the cell gains a responsiveness to a signal

THE AUTHOR

TERRY KENAKIN has been applying concepts of receptor pharmacology to drug discovery programs for almost three decades, most recently as a principal research investigator at the pharmaceutical firm GlaxoSmithKline. He has written six books on pharmacology and is co-editor in chief of the *Journal of Receptors and Signal Transduction*.

it would otherwise ignore. Individual proteins have their blueprints in specific genes, but these combination receptors have no corresponding single blueprint (from which their behaviors might be predicted), so they might be thought of as products of "phantom" genes.

In some cases, the novel receptor is a complex consisting of two or more GPCRs. In other cases, it consists of a GPCR and a co-protein—one that is not itself a receptor but gives the receptor an altered set of properties. The receptor for a hormone called amylin seems to be

of this type. Released by the same pancreatic cells that secrete insulin, amylin modulates the effects of insulin on other cells, but efforts to identify a single protein that serves as its receptor have failed. What is more, analyses of the recently completed human genome se-

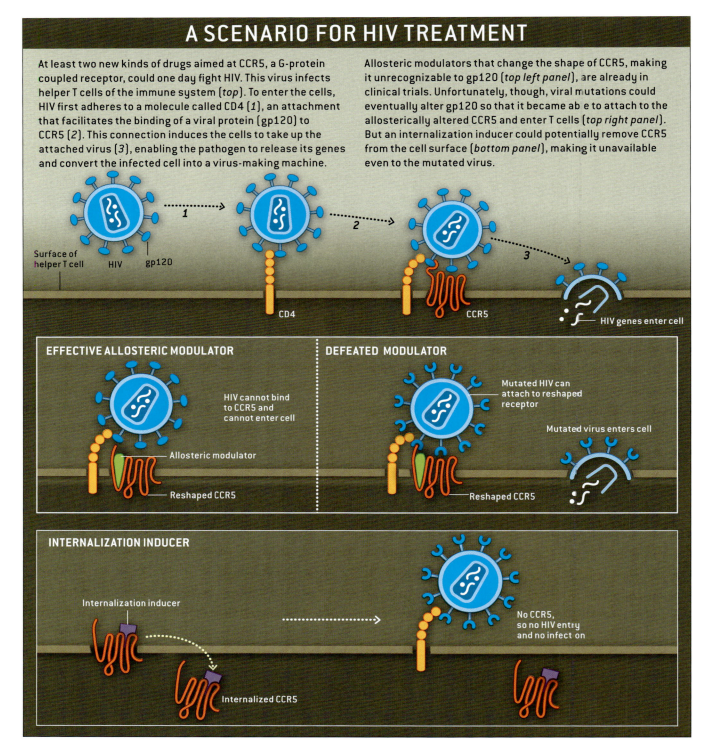

A SCENARIO FOR HIV TREATMENT

At least two new kinds of drugs aimed at CCR5, a G-protein coupled receptor, could one day fight HIV. This virus infects helper T cells of the immune system (*top*). To enter the cells, HIV first adheres to a molecule called CD4 (*1*), an attachment that facilitates the binding of a viral protein (gp120) to CCR5 (*2*). This connection induces the cells to take up the attached virus (*3*), enabling the pathogen to release its genes and convert the infected cell into a virus-making machine.

Allosteric modulators that change the shape of CCR5, making it unrecognizable to gp120 (*top left panel*), are already in clinical trials. Unfortunately, though, viral mutations could eventually alter gp120 so that it became able to attach to the allosterically altered CCR5 and enter T cells (*top right panel*). But an internalization inducer could potentially remove CCR5 from the cell surface (*bottom panel*), making it unavailable even to the mutated virus.

Surface of helper T cell   HIV   gp120

1   2   3

CD4   CCR5   HIV genes enter cell

EFFECTIVE ALLOSTERIC MODULATOR

HIV cannot bind to CCR5 and cannot enter cell

Allosteric modulator

Reshaped CCR5

DEFEATED MODULATOR

Mutated HIV can attach to reshaped receptor

Mutated virus enters cell

Reshaped CCR5

INTERNALIZATION INDUCER

Internalization inducer

Internalized CCR5

No CCR5, so no HIV entry and no infection

SOME EARLY PROSPECTS FOR NEW DRUGS

For the most part, investigators are only beginning to devise drugs that influence GPCRs in new ways. But many such agents can be expected to enter pharmaceutical pipelines in the years ahead.

DISORDER	DRUG TYPE	DRUG NAME (MAKER)	TARGET GPCR	STAGE OF DEVELOPMENT
HIV infection	Allosteric modulator	Aplaviroc (GlaxoSmithKline); Vicriviroc (Schering-Plough); UK-427, 857 (Pfizer)	CCR5 (binding by HIV helps the virus enter cells)	All are in phase II or III human trials (early or advanced tests of efficacy)
	Allosteric modulator	AMD3100 (AnorMED)	CXCR4 (this receptor, too, can help HIV enter cells)	In phase III human trials
	Internalization inducer	PSC-RANTES (several institutions)	CCR5	Theoretical
Diabetes	Binder of a receptor formed by two molecules	Symlin (Amylin)	Complex consisting of a protein called RAMP and the GPCR for calcitonin (a thyroid hormone)	Gained U.S. approval in March 2005
Obesity	Inverse agonist	None yet	Constitutively active ghrelin receptor in central nervous system	Theoretical
	Inverse agonist	None yet	Constitutively active histamine H_3 receptor in central nervous system	Theoretical
Cancer	Inverse agonist	None yet	Various constitutively active GPCRs	Theoretical

quence indicate that no gene for such a receptor exists. On the other hand, a complex consisting of the GPCR for the thyroid hormone calcitonin plus a nonreceptor protein called RAMP (receptor activity–modifying protein) responds strongly and selectively to amylin. Apparently RAMP makes the calcitonin receptor "multilingual"—that is, the receptor is reactive to calcitonin if cells lack RAMP, but it is sensitive to amylin if cells contain RAMP.

A different co-protein, called RCP (receptor component protein), induces the calcitonin receptor to obey signals from yet another substance—CGRP (calcitonin-gene-related peptide), a small protein that is the most potent known dilator of blood vessels. This conversion becomes valuable during pregnancy, when blood levels of the dilating peptide soar and RCP levels rise in the uterine wall. As RCP concentrations increase, so do the numbers of calcitonin receptors that become sensitive to the dilator, a change that enhances the blood supply to tissues important for childbirth.

Because co-proteins affect GPCR activity, they might themselves prove

valuable as drug targets. One intriguing target is modulin, a co-protein that binds to receptors for serotonin. In the brain, serotonin is most famous as a mood-enhancing neurotransmitter. (Prozac and related antidepressants work by increasing the brain's serotonin levels.) Outside the brain, it acts on the intestines and blood vessels. Perhaps unsurprisingly, serotonin receptors have numerous subtypes, and modulin further tunes the effects of serotonin on particular cells by altering a subtype's sensitivity to it. A drug that mimicked or inhibited modulin, then, could in theory increase or decrease the responsiveness of specific serotonin receptors on specific cell types and might thereby be beneficial in realms ranging from schizo-

phrenia to gastrointestinal function.

Researchers estimate that of the estimated 650 human GPCR genes, about 330 might be blueprints for receptors well worth targeting by drugs. In the past, pharmaceutical scientists would have focused strictly on developing old-fashioned inhibitors or agonists aimed at the receptors' active site. But if many GPCRs offer multiple sites of attack, the opportunities for devising new therapies explode. Because it can take 15 or even 20 years to discover a drug, explore its actions, evaluate its safety, and get it to market, detailed forecasts would be premature. Nevertheless, the new insights into how GPCRs are controlled suggest that these old standbys still have exciting tales to tell.   SA

MORE TO EXPLORE

Novel GPCRs and Their Endogenous Ligands: Expanding the Boundaries of Physiology and Pharmacology. A. Marchese, S. R. George, L. F. Kolakowski, K. R. Lynch and B. F. O'Dowd in Trends in Pharmacological Sciences, Vol. 20, No. 9, pages 370–375; September 1, 1999.

Drug Discovery: A Historical Perspective. J. Drews in Science, Vol. 287, pages 1960–1964; March 17, 2000.

G-Protein-Coupled Receptor Interacting Proteins: Emerging Roles in Localization and Signal Transduction. A. E. Brady and L. E. Limbird in Cellular Signalling, Vol. 14, No. 4, pages 297–309; April 2002.

A Pharmacology Primer: Theory, Application, and Methods. Terry Kenakin. Academic Press (Elsevier), 2003.

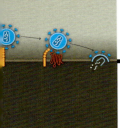

Article Review: CELL BIOLOGY

New Bulls-eye for Drugs — *by Terry Kenakin*

TESTING YOUR COMPREHENSION

1. Define what allosteric regulation of a protein involves. How do pharmaceutically-produced allosteric modulators of GPCRs work?

   ANSWER: *Allosteric effects upon a protein identify structural changes occurring in that protein that usually affect its enzymatic activity. The allosteric regulation (activation, inhibition) of a protein indicates interactions with a regulatory molecule, binding outside of the enzyme active site, that change the proteins conformation to inhibit or activate the enzyme. Allosteric modulators bind to the GPCR (outside of ligand-binding site), generating a conformational change that inhibits binding by the ligand.*

2. Define two unique signaling phenomenons that can be stimulated by norepinephrine.

   ANSWER: *Norepinephrine has varied signaling capacities that include affecting the heart rate, dilation air passages, and blood pressure.*

3. Define the terms agonist and antagonist, as they relate to receptors and ligands.

   ANSWER: *An agonist is any ligand (receptor binding molecule) that alters the activity of the receptor. Generally agonists increase the activity of the receptor. An antagonist reduces or interferes with the agonist's ability to regulate receptor activity.*

4. GPCRs are no longer considered "toggle" switches that activate a pathway in response to a ligand-based molecular signal, and inactivate the pathway when the ligand concentration decreases. Explain the newer model for GPCR signaling.

   ANSWER: *An individual GPCR can often bind to a number of different G-proteins. There are four classes of G-proteins, encompassing a wide array of signaling capacities. The impact each GPCR has depends upon the G-proteins that associate with that GPCR, and their relative levels of expression in a given cell type.*

CLASS ACTIVITIES AND DISCUSSION

1. Examine the table listing *marketed* GPCR-targeted pharmaceuticals. Divide the class into groups of 3-4 students, have each group choose one of the drugs on the list, and then research that drug in detail. Key questions to be addressed should include: the specific GPCR protein(s) targeted by that drug, the biological process that is regulated by that specific GPCR, the nature of abnormal signaling by that GPCR that generates the disorder being treated, and listed side affects. Do the side affects make sense from what you know about the cellular processes impacted by the GPCR?

2. The "G" in G-protein signaling indicates a heterotrimeric GTP-binding protein complex that associates with the receptor on the cytosolic side of the membrane. The ligand-binding site for GPCR proteins is found on the outside of the cell. How does a receptor transmit the "ligand bound" signal from the outside of the cell to the inside of the cell? Develop a list of mechanisms that mediate this process of signal transmission to the G-protein.

Article Review: GENETICS

New Bulls-eye for Drugs **by Terry Kenakin**

TESTING YOUR COMPREHENSION

1. Amylin is a hormone produced by pancreatic cells and modulates the affect of insulin upon various cell types. Amylin exerts its influence through a GPCR that is not encoded by a single gene. Explain this unusual phenomenon.

   ANSWER: *Amylin binds to a GPCR that is composed of the calcitonin GPCR and a co-protein called RAMP (receptor-activity-modifying protein). The two proteins together generate an amylin responsive GPCR protein that signals amylin binding via G-protein signaling. Also, notice that the calcitonin GPCR can bind a second co-protein (RCP) generating an entirely different binding and signaling profile.*

2. In an initial examination of the human genome sequence, how many GPCR genes were found?

   ANSWER: *~650 are predicted by analysis of the DNA sequence alone.*

3. Explain the role modulin protein plays in GPCR signaling.

   ANSWER: *Modulin is a co-protein that associates with the serotonin GPCR receptor. Modulin binding with this GPCR may alter the downstream signaling affects of serotonin signaling at the cell surface. Modulin may serve as a target for drug-based treatment of diseases that are thought to originate with abnormal serotonin signaling.*

4. Norepinephrine is a neurotransmitter that is capable of binding two classes of GPCR (alpha and beta adrenoceptors) to stimulate their distinct signaling activities. How do these unique classes of GPCR bind to the same ligand, and yet direct very different cellular signals?

   ANSWER: *GPCR genes and G-protein genes are expressed differently in different tissue types. Thus each GPCR will interact with available G-proteins, depending upon the affinity of that GPCR for the G-proteins. Binding ligand will activate the GPCR, to activate signaling from the associated G-protein. Different G-proteins ultimately will impact the cell in different ways.*

CLASS ACTIVITIES AND DISCUSSION

1. CCR5 is a GPCR that mediates inflammatory signaling in cells of the immune system. If a drug that modifies CCR5 structure (to inhibit binding to gp140) also impacts CCR5 signaling, what impact might that have upon the body? If you were a research scientist studying the impact of a specific CCR5 allosteric modulator upon HIV infection, how might you assess the impact of your drug upon the immune system?

2. Alleles of a mutant form of CCR5 (called Δ32) are found in the human population. Perform a PubMed search to determine the nature of this mutation. Is the mutant protein functional? How does this mutation confer resistance to HIV entry into cells?

3. Compare the impact of the CCR5-Δ32 mutation upon HIV infection to the affects of CCR5 allosteric modulators and CCR5 internalization inducers. In this case, why are internalization inducers thought to be superior to allosteric modulators?

4. Use online resources to research *McCune–Albright syndrome*. What gene is mutated in individuals suffering from this disease? What is the function of the encoded protein? Identify the tissues affected, and describe the symptoms of the syndrome in detail. This disease is caused by spontaneous mutation in a gene, and generates individuals who are mosaic for the disease trait. Individuals with this mutation in all of their cells have not been observed, why do you think this is?

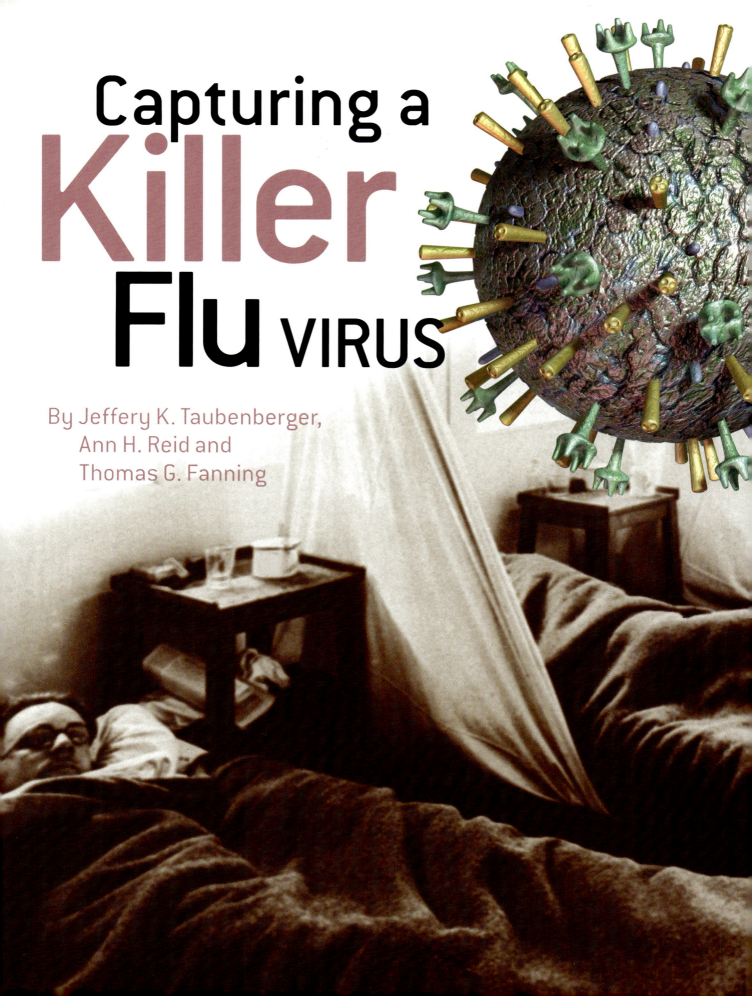

Capturing a
Killer
Flu VIRUS

By Jeffery K. Taubenberger,
Ann H. Reid and
Thomas G. Fanning

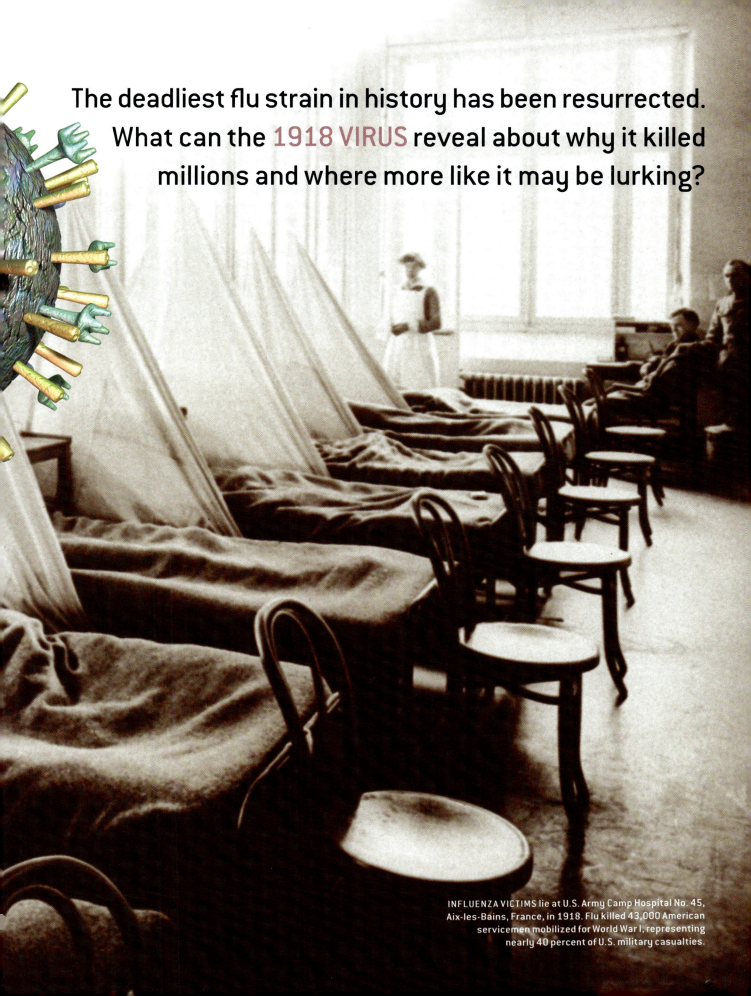

The deadliest flu strain in history has been resurrected. What can the 1918 VIRUS reveal about why it killed millions and where more like it may be lurking?

INFLUENZA VICTIMS lie at U.S. Army Camp Hospital No. 45, Aix-les-Bains, France, in 1918. Flu killed 43,000 American servicemen mobilized for World War I, representing nearly 40 percent of U.S. military casualties.

O n September 7, 1918, at the height of World War I, a soldier at an army training camp outside Boston came to sick call with a high fever. Doctors diagnosed him with meningitis but changed their minds the next day when a dozen more soldiers were hospitalized with respiratory symptoms. Thirty-six new cases of this unknown illness appeared on the 16th. Incredibly, by September 23rd, 12,604 cases had been reported in the camp of 45,000 soldiers. By the end of the outbreak, one third of the camp's population would come down with this severe disease, and nearly 800 of them would die. The soldiers who perished often developed a bluish skin color and struggled horribly before succumbing to death by suffocation. Many died less than 48 hours after their symptoms appeared, and at autopsy their lungs were filled with fluid or blood.

Because this unusual suite of symptoms did not fit any known malady, a distinguished pathologist of the era, William Henry Welch, speculated that "this must be some new kind of infection or plague." Yet the disease was neither plague nor even new. It was just influenza. Still, this particularly virulent and infectious strain of the flu virus is thought to have killed as many as 40 million people around the world between 1918 and 1919.

This most lethal flu outbreak in modern history disappeared almost as quickly as it emerged, and its cause was long believed lost to time. No one had preserved samples of the pathogen for later

RED CROSS NURSES in St. Louis carry a flu patient in 1918. Health workers, police and a panicked public donned face masks for protection as the virus swept the country. Nearly a third of all Americans were infected during the pandemic, and 675,000 of them died.

study because influenza would not be identified as a virus until the 1930s. But thanks to incredible foresight by the U.S. Army Medical Museum, the persistence of a pathologist named Johan Hultin, and advances in genetic analysis of old tissue samples, we have been able to retrieve parts of the 1918 virus and study their features. Now, more than 80 years after the horrible natural disaster of 1918–1919, tissues recovered from a handful of victims are answering fundamental questions both about the nature of this pandemic strain and about the workings of influenza viruses in general.

The effort is not motivated merely by historical curiosity. Because influenza viruses continually evolve, new influenza strains continually threaten human populations. Pandemic human flu virus-es have emerged twice since 1918—in 1957 and 1968. And flu strains that usually infect only animals have also periodically caused disease in humans, as seen in the recent outbreak of avian influenza in Asia. Our two principal goals are determining what made the 1918 influenza so virulent, to guide development of influenza treatments and preventive measures, and establishing the origin of the pandemic virus, to better target possible sources of future pandemic strains.

Hunting the 1918 Virus

IN MANY RESPECTS, the 1918 influenza pandemic was similar to others before it and since. Whenever a new flu strain emerges with features that have never been encountered by most people's immune systems, widespread flu outbreaks are likely. But certain unique characteristics of the 1918 pandemic have long remained enigmatic.

For instance, it was exceptional in both its breadth and depth. Outbreaks swept across Europe and North America, spreading as far as the Alaskan wilderness and the most remote islands of the Pacific. Ultimately, one third of the world's population may have been infected. The disease was also unusually severe, with death rates of 2.5 to 5 percent—up to 50 times the mortality seen in other influenza outbreaks.

Overview/ *The Mystery of 1918*

- The flu pandemic that swept the globe in 1918–1919 was exceptional for the sheer numbers it killed, especially the number of young people who succumbed to the unusually virulent flu virus.
- What made the strain so deadly was a longstanding medical mystery until the authors devised techniques that allowed them to retrieve the 1918 virus's genes from victims' preserved tissues.
- Analysis of those genes and the proteins they encode revealed viral features that could have both suppressed immune defenses and provoked a violent immune reaction in victims, contributing to the high mortality.
- Known bird and mammal influenza hosts are unlikely sources of the pandemic virus, so its origin remains unsolved.

By the fall of 1918 everyone in Europe was calling the disease the "Spanish" influenza, probably because neutral Spain did not impose the wartime censorship of news about the outbreak prevalent in combatant countries. The name stuck, although the first outbreaks, or spring wave, of the pandemic seemingly arose in and around military camps in the U.S. in March 1918. The second, main wave of the global pandemic occurred from September to November 1918, and in many places yet another severe wave of influenza hit in early 1919.

Antibiotics had yet to be discovered, and most of the people who died during the pandemic succumbed to pneumonia caused by opportunistic bacteria that infected those already weakened by the flu. But a subset of influenza victims died just days after the onset of their symp-

the 1918 virus preserved in the victims' lungs. Unfortunately, all attempts to culture live influenza virus from these specimens were unsuccessful.

In 1995 our group initiated an attempt to find the 1918 virus using a different source of tissue: archival autopsy specimens stored at the Armed Forces Institute of Pathology (AFIP). For several years, we had been developing expertise in extracting fragile viral genetic material from damaged or decayed tissue for diagnostic purposes. In 1994, for instance, we were able to use our new techniques to help an AFIP marine mammal pathologist investigate a mass dolphin die-off that had been blamed on red tide. Although the available dolphin tissue samples were badly decayed, we extracted enough pieces of RNA from them to identify a new virus, similar to the one that

damage characteristic of patients who died rapidly. Because the influenza virus normally clears the lungs just days after infection, we had the greatest chance of finding virus remnants in these victims.

The standard practice of the era was to preserve autopsy specimens in formaldehyde and then embed them in paraffin, so fishing out tiny genetic fragments of the virus from these 80-year-old "fixed" tissues pushed the very limits of the techniques we had developed. After an agonizing year of negative results, we found the first influenza-positive sample in 1996, a lung specimen from a soldier who died in September 1918 at Fort Jackson, S.C. We were able to determine the sequence of nucleotides in small fragments of five influenza genes from this sample.

But to confirm that the sequences belonged to the lethal 1918 virus, we kept

After an AGONIZING YEAR of negative results, we found THE FIRST CASE in 1996.

toms from a more severe viral pneumonia—caused by the flu itself—that left their lungs either massively hemorrhaged or filled with fluid. Furthermore, most deaths occurred among young adults between 15 and 35 years old, a group that rarely dies from influenza. Strikingly, people younger than 65 years accounted for more than 99 percent of all "excess" influenza deaths (those above normal annual averages) in 1918–1919.

Efforts to understand the cause of the 1918 pandemic and its unusual features began almost as soon as it was over, but the culprit virus itself remained hidden for nearly eight decades. In 1951 scientists from the University of Iowa, including a graduate student recently arrived from Sweden named Johan Hultin, went as far as the Seward Peninsula of Alaska seeking the 1918 strain [see box on page 31]. In November 1918 flu spread through an Inuit fishing village now called Brevig Mission in five days, killing 72 people—about 85 percent of the adult population. Their bodies had since been buried in permafrost, and the 1951 expedition members hoped to find

causes canine distemper, which proved to be the real cause of the dolphin deaths. Soon we began to wonder if there were any older medical mysteries we might solve with our institute's resources.

A descendant of the U.S. Army Medical Museum founded in 1862, the AFIP has grown along with the medical specialty of pathology and now has a collection of three million specimens. When we realized that these included autopsy samples from 1918 flu victims, we decided to go after the pandemic virus. Our initial study examined 78 tissue samples from victims of the deadly fall wave of 1918, focusing on those with the severe lung

looking for more positive cases and identified another one in 1997. This soldier also died in September 1918, at Camp Upton, N.Y. Having a second sample allowed us to confirm the gene sequences we had, but the tiny quantity of tissue remaining from these autopsies made us worry that we would never be able to generate a complete virus sequence.

A solution to our problem came from an unexpected source in 1997: Johan Hultin, by then a 73-year-old retired pathologist, had read about our initial results. He offered to return to Brevig Mission to try another exhumation of 1918 flu victims interred in permafrost. Forty-

THE AUTHORS

JEFFERY K. TAUBENBERGER, ANN H. REID and THOMAS G. FANNING work together at the Armed Forces Institute of Pathology in Rockville, Md. In 1993 Taubenberger, a molecular pathologist, helped to create a laboratory there devoted to molecular diagnostics—identifying diseases by their genetic signatures rather than by the microscopic appearance of patients' tissue samples. Early work by Reid, a molecular biologist, led the group to devise the techniques for extracting DNA and RNA from damaged or decayed tissue that allowed them to retrieve bits and pieces of 1918 flu virus genes from archived autopsy specimens. Fanning, a geneticist with expertise in the evolution of genomes, helped to analyze the genes' relationships to other animal and human flu viruses. The authors wish to note that the opinions expressed in this article are their own and do not represent the views of the Department of Defense or the AFIP.

Influenza is a small and simple virus—just a hollow lipid ball studded with a few proteins and bearing only eight gene segments (*below*). But that is all it needs to induce the cells of living hosts to make more viruses (*bottom*). One especially important protein on influenza's surface, hemagglutinin (HA), allows the virus to enter cells. Its shape determines which hosts a flu virus strain can infect. Another protein, neuraminidase (NA), cuts newly formed viruses loose from an infected cell, influencing how efficiently the virus can spread. Slight changes in these and other flu proteins can help the virus infect new kinds of hosts and evade immune attack. The alterations can arise through mistakes that occur while viral genes are being copied. Or they can be acquired in trade when the genes of two different flu viruses infecting the same cell intermingle (*right*).

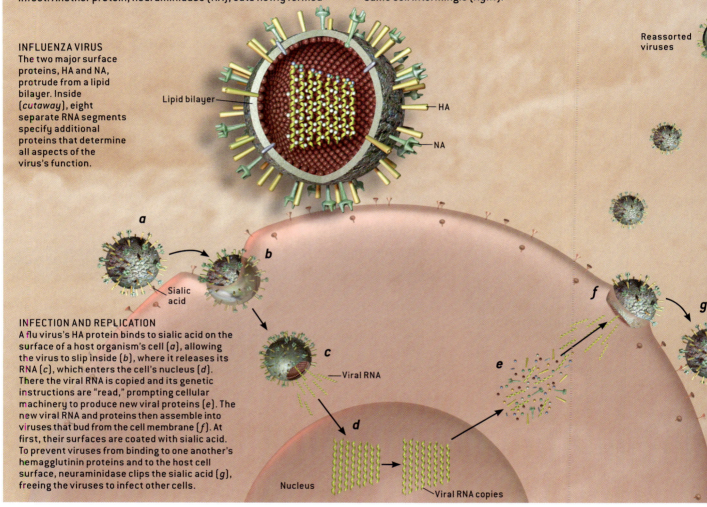

Reassorted viruses

INFLUENZA VIRUS
The two major surface proteins, HA and NA, protrude from a lipid bilayer. Inside (*cutaway*), eight separate RNA segments specify additional proteins that determine all aspects of the virus's function.

Lipid bilayer

HA

NA

a

b

Sialic acid

INFECTION AND REPLICATION
A flu virus's HA protein binds to sialic acid on the surface of a host organism's cell (*a*), allowing the virus to slip inside (*b*), where it releases its RNA (*c*), which enters the cell's nucleus (*d*). There the viral RNA is copied and its genetic instructions are "read," prompting cellular machinery to produce new viral proteins (*e*). The new viral RNA and proteins then assemble into viruses that bud from the cell membrane (*f*). At first, their surfaces are coated with sialic acid. To prevent viruses from binding to one another's hemagglutinin proteins and to the host cell surface, neuraminidase clips the sialic acid (*g*), freeing the viruses to infect other cells.

c

Viral RNA

d

e

f

g

Nucleus

Viral RNA copies

six years after his first attempt, with permission from the Brevig Mission Council, he obtained frozen lung biopsies of four flu victims. In one of these samples, from a woman of unknown age, we found influenza RNA that provided the key to sequencing the entire genome of the 1918 virus.

More recently, our group, in collaboration with British colleagues, has also been surveying autopsy tissue samples from 1918 influenza victims from the Royal London Hospital. We have been able to analyze flu virus genes from two of these cases and have found that they were nearly identical to the North American samples, confirming the rapid worldwide spread of a uniform virus. But what can the sequences tell us about the virulence and origin of the 1918 strain? Answering those questions requires a bit of background about how influenza viruses function and cause disease in different hosts.

Flu's Changing Face

EACH OF THE THREE novel influenza strains that caused pandemics in the past 100 years belonged to the type A group of flu viruses. Flu comes in three main forms, designated A, B and C. The latter two infect only humans and have never caused pandemics. Type A influenza viruses, on the other hand, have been found to infect a wide variety of animals, including poultry, swine, horses, humans and other mammals. Aquatic birds, such as ducks, serve as the natural "reservoir" for all the known subtypes of influenza A, meaning that the virus infects the bird's gut without causing symptoms. But these wild avian strains

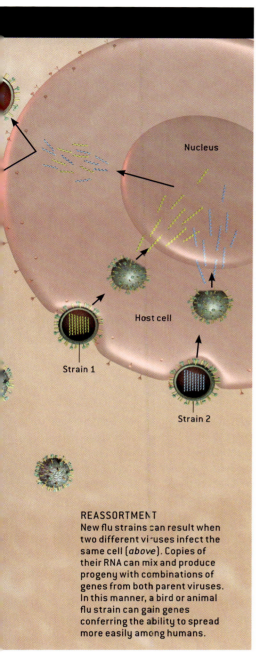

REASSORTMENT
New flu strains can result when two different viruses infect the same cell (*above*). Copies of their RNA can mix and produce progeny with combinations of genes from both parent viruses. In this manner, a bird or animal flu strain can gain genes conferring the ability to spread more easily among humans.

Within the image: Nucleus, Host cell, Strain 1, Strain 2

can mutate over time or exchange genetic material with other influenza strains, producing novel viruses that are able to spread among mammals and domestic poultry.

The life cycle and genomic structure of influenza A virus allow it to evolve and exchange genes easily. The virus's genetic material consists of eight separate RNA segments encased in a lipid membrane studded with proteins [*see top illustration on opposite page*]. To reproduce, the virus binds to and then enters a living cell, where it commandeers cel-

lular machinery, inducing it to manufacture new viral proteins and additional copies of viral RNA. These pieces then assemble themselves into new viruses that escape the host cell, proceeding to infect other cells. No proofreading mechanism ensures that the RNA copies are accurate, so mistakes leading to new mutations are common. What is more, should two different influenza virus strains infect the same cell, their RNA segments can mix freely there, producing progeny viruses that contain a combination of genes from both the original viruses. This "reassortment" of viral genes is an important mechanism for generating diverse new strains.

Different circulating influenza A viruses are identified by referring to two signature proteins on their surfaces. One is hemagglutinin (HA), which has at least 15 known variants, or subtypes. Another is neuraminidase (NA), which has nine subtypes. Exposure to these proteins produces distinctive antibodies in a host, thus the 1918 strain was the first to be named, "H1N1," based on antibodies found in the bloodstream of pandemic survivors. Indeed, less virulent descendants of H1N1 were the predominant circulating flu strains until 1957, when an H2N2 virus emerged, causing a pandemic. Since 1968, the H3N2 subtype, which provoked the pandemic that year, has predominated.

The HA and NA protein subtypes present on a given influenza A virus are more than just identifiers; they are essential for viral reproduction and are primary targets of an infected host's immune system. The HA molecule initiates infection by binding to receptors on the surface of certain host cells. These tend to be respiratory lining cells in mammals and intestinal lining cells in birds. The NA protein enables new virus copies to escape the host cell so they can go on to infect other cells.

After a host's first exposure to an HA subtype, antibodies will block receptor binding in the future and are thus very effective at preventing reinfection with the same strain. Yet flu viruses with HA subtypes that are new to humans periodically appear, most likely through

reassortment with the extensive pool of influenza viruses infecting wild birds. Normally, influenza HAs that are adapted to avian hosts bind poorly to the cell-surface receptors prevalent in the human respiratory tract, so an avian virus's HA binding affinity must be somewhat modified before the virus can replicate and spread efficiently in humans. Until recently, existing evidence suggested that a wholly avian influenza virus probably could not directly infect humans, but 18 people were infected with an avian H5N1 influenza virus in Hong Kong in 1997, and six died.

Outbreaks of an even more pathogenic version of that H5N1 strain became widespread in Asian poultry in 2003 and 2004, and more than 30 people infected with this virus have died in Vietnam and Thailand.

The virulence of an influenza virus once it infects a host is determined by a complex set of factors, including how readily the virus enters different tissues, how quickly it replicates, and the violence of the host's immune response to the intruder. Thus, understanding exactly what made the 1918 pandemic influenza strain so infectious and so virulent could yield great insight into what makes any influenza strain more or less of a threat.

A Killer's Face

WITH THE 1918 RNA we have retrieved, we have used the virus's own genes as recipes for manufacturing its component parts—essentially re-creating pieces of the killer virus itself. The first of these we were eager to examine was the hemagglutinin protein, to look for features that might explain the exceptional virulence of the 1918 strain.

We could see, for example, that the part of the 1918 HA that binds with a host cell is nearly identical to the binding site of a wholly avian influenza HA [*see illustration on page 29*]. In two of the 1918 isolates, this receptor-binding site differs from an avian form by only one amino acid building block. In the other three isolates, a second amino acid is also altered. These seemingly subtle mutations may represent the min-

When analyzing the genes of the 1918 virus revealed no definitive reasons for the pandemic strain's virulence, our group turned to reverse genetics—a method of understanding the function of genes by studying the proteins they encode. In collaboration with scientists from the Mount Sinai School of Medicine, the Centers for Disease Control and Prevention, the U.S. Department of Agriculture, the University of Washington and the Scripps Research Institute, we "built" influenza viruses containing one or more of the 1918 virus's genes, so we could see how these recombinant viruses behaved in animals and human cell cultures.

To construct these viruses, we employed a new technique called plasmid-based reverse genetics, which requires first making DNA copies of flu genes that normally exist in RNA form. Each DNA gene copy is then inserted into a tiny ring of DNA called a plasmid. Different combinations of these plasmids can be injected into living cells, where cellular machinery will execute the genetic instructions they bear and manufacture flu viruses with only the desired combination of genes.

Reverse genetics not only allows us to study the 1918 virus, it will allow scientists in the U.S. and Europe to explore how great a threat the H5N1 avian flu virus poses to humans. Since January 2004, that strain—which is now present in birds in 10 Asian countries—has infected more than 40 people, killing more than 30 of them. One of the casualties was a mother who is believed to have contracted the virus from her daughter, rather than directly from a bird.

Such human-to-human transmission could suggest that in their case the avian virus had adapted to be more easily spread between humans, either by mutating or by acquiring new genes through reassortment with a circulating human flu strain. That dreaded development would increase the possibility of a human pandemic. Hoping to predict and thereby prevent such a disaster, scientists at the CDC and Erasmus University in the Netherlands are planning to test combinations of H5N1 with current human flu strains to assess the likelihood of their occurring naturally and their virulence in people.

What these experiments will reveal, as in our group's work with the 1918 virus genes, is crucial to understanding how influenza pandemics form and why they cause disease. Some observers have questioned the safety of experimenting with lethal flu strains, but all of this research is conducted in secure laboratories designed specifically to deal with highly pathogenic influenza viruses.

What is more, re-creating the 1918 virus proteins enabled us to establish that currently available antiviral drugs, such as amantadine or the newer neuraminidase inhibitors, such as oseltamivir (Tamiflu), would be effective against the 1918 strain in the case of an accidental infection. The H5N1 viruses are also sensitive to the neuraminidase inhibitors.

Scientists in the U.S. and U.K. also recently employed plasmid-based reverse genetics to create a seed strain for a human vaccine against H5N1. They made a version of the H5N1 virus lacking the wild strain's most deadly features, so that manufacturers could safely use it to produce a vaccine [see "The Scientific American 50," December 2004]. Clinical trials of that H5N1 vaccine were scheduled to begin at the end of 2004.

—J.K.T., A.H.R. and T.G.F.

PLASMID-BASED reverse genetics lets scientists custom-manufacture flu viruses. DNA copies of genes from two different flu strains (*blue* and *red*) are inserted into DNA rings called plasmids. The gene-bearing plasmids are then injected into a culture of living cells, which manufacture whole flu viruses containing the desired combination of genes.

Gene copies

Plasmids

Cell culture

New flu virus

imal change necessary to allow an avian-type HA to bind to mammalian-type receptors.

But while gaining a new binding affinity is a critical step that allows a virus to infect a new type of host, it does not necessarily explain why the 1918 strain was so lethal. We turned to the gene sequences themselves, looking for features that could be directly related to virulence, including two known mutations in other flu viruses. One involves the HA gene: to become active in a cell, the HA protein must be cleaved into two pieces by a gut-specific protein-cutting enzyme, or protease, supplied by the host. Some avian H5 and H7 subtype viruses acquire a gene mutation that adds one or more basic amino acids to the cleavage site, allowing HA to be activated by ubiquitous proteases. In chickens and other birds, infection by such a virus causes disease in multiple organs and even the central nervous system, with a very high mortality rate. This mutation has been observed in the H5N1 viruses currently circulating in Asia. We did not, however, find it in the 1918 virus.

The other mutation with a significant effect on virulence has been seen in the NA gene of two influenza virus strains that infect mice. Again, mutations at a single amino acid appear to allow the virus to replicate in many different body tissues, and these flu strains are typically lethal in laboratory mice. But we did not see this mutation in the NA of the 1918 virus either.

Because analysis of the 1918 virus's

genes was not revealing any characteristics that would explain its extreme virulence, we initiated a collaborative effort with several other institutions to re-create parts of the 1918 virus itself so we could observe their effects in living tissues.

A new technique called plasmid-based reverse genetics allows us to copy 1918 viral genes and then combine them with the genes of an existing influenza strain, producing a hybrid virus. Thus, we can take an influenza strain adapted to mice, for example, and give it different combinations of 1918 viral genes. Then, by infecting a live animal or a human tissue culture with this engineered virus, we can see which components of

a tissue culture of human lung cells, we found that a virus with the 1918 NS1 gene was indeed more effective at blocking the host's type I IFN system.

To date, we have produced recombinant influenza viruses containing between one and five of the 1918 genes. Interestingly, we found that any of the recombinant viruses possessing both the 1918 HA and NA genes were lethal in mice, causing severe lung damage similar to that seen in some of the pandemic fatalities. When we analyzed these lung tissues, we found signatures of gene activation involved in common inflammatory responses. But we also found higher than normal activation of

unclear, this protein may have played a key role in the 1918 strain's virulence.

These ongoing experiments are providing a window to the past, helping scientists understand the unusual characteristics of the 1918 pandemic. Similarly, these techniques will be used to study what types of changes to the current H5N1 avian influenza strain might give that extremely lethal virus the potential to become pandemic in humans [*see box on opposite page*]. An equally compelling question is how such virulent strains emerge in the first place, so our group has also been analyzing the 1918 virus's genes for clues about where it might have originated.

Seemingly subtle mutations may allow an AVIAN hemagglutinin to bind to MAMMALIAN receptors.

the pandemic strain might have been key to its pathogenicity.

For instance, the 1918 virus's distinctive ability to produce rapid and extensive damage to both upper and lower respiratory tissues suggests that it replicated to high numbers and spread quickly from cell to cell. The viral protein NS1 is known to prevent production of type I interferon (IFN)—an "early warning" system that cells use to initiate an immune response against a viral infection. When we tested recombinant viruses in

genes associated with the immune system's offensive soldiers, T cells and macrophages, as well as genes related to tissue injury, oxidative damage, and apoptosis, or cell suicide.

More recently, Yoshihiro Kawaoka of the University of Wisconsin–Madison reported similar experiments with 1918 flu genes in mice, with similar results. But when he tested the HA and NA genes separately, he found that only the 1918 HA produced the intensive immune response, suggesting that for reasons as yet

Seeking the Source

THE BEST APPROACH to analyzing the relationships among influenza viruses is phylogenetics, whereby hypothetical family trees are constructed using viral gene sequences and knowledge of how often genes typically mutate. Because the genome of an influenza virus consists of eight discrete RNA segments that can move independently by reassortment, these evolutionary studies must be performed separately for each gene segment.

We have completed analyses of five of the 1918 virus's eight RNA segments, and so far our comparisons of the 1918 flu genes with those of numerous human, swine and avian influenza viruses always place the 1918 virus within the human and swine families, outside the avian virus group [*see box on next page*]. The 1918 viral genes do have some avian features, however, so it is probable that the virus originally emerged from an avian reservoir sometime before 1918. Clearly by 1918, though, the virus had acquired enough adaptations to mammals to function as a human pandemic virus. The question is, where?

When we analyzed the 1918 hemagglutinin gene, we found that the sequence has many more differences from

Human-adapted   H3

Avian-adapted   H5

1918 Flu   H1

Amino acid change

HEMAGGLUTININ (HA) of the 1918 flu strain was re-created from its gene sequence by the authors' collaborators so they could examine the part that binds to a host cell's sialic acid and allows the virus to enter the cell. HA binding sites usually are shaped differently enough to bar cross-species infection. For instance, the human-adapted H3-type HA has a wide cavity in the middle of its binding site (*left*), whereas the avian H5 cavity (*center*) is narrow. The 1918 H1-type HA (*right*) more closely resembles the avian form, with only a few minor differences in the sequence of its amino acid building blocks. One of these alterations (*above right*) slightly widens the central cavity, apparently just enough to have allowed a flu virus with this avian-type HA to infect hundreds of millions of humans in 1918–1919.

Flu Family Tree

Seeking clues to the origin of the 1918 virus's hemagglutinin (HA), the authors analyzed gene sequences for the H1-subtype of HA from a variety of flu strains and constructed a phylogeny showing their evolutionary relationships. Samples of the 1918 strain (S. Carolina, New York, Brevig) fell within the family of human-adapted flu viruses. The 1918 H1 gene's distance from the known avian family could indicate that it originated in an avian flu strain but spent time evolving in an unidentified host before emerging in 1918. Supporting this conclusion, a contemporary avian strain found in a preserved Brant goose (Alaska 1917) was evolutionarily distant from the 1918 strain and more similar to modern bird flus.

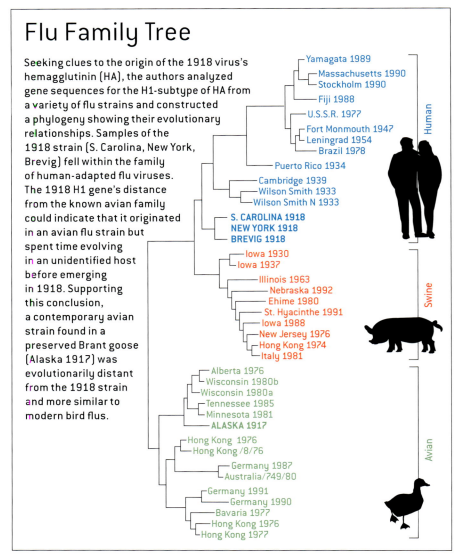

mans and swine during the 1918 pandemic, but we believe that the direction of transmission was most probably from humans to pigs. There are numerous examples of human influenza A virus strains infecting swine since 1918, but swine influenza strains have been isolated only sporadically from humans. Nevertheless, to explore the possibility that the 1918 HA may have started as an avian form that gradually adapted to mammalian hosts in swine, we looked at a current example of how avian viruses evolve in pigs—an avian H1N1 influenza lineage that has become established in European swine over the past 25 years. We found that even 20 years of evolution in swine has not resulted in the number of changes from avian sequences exhibited by the 1918 pandemic strain.

When we applied these types of analyses to four other 1918 virus genes, we came to the same conclusion: the virus that sparked the 1918 pandemic could well have been an avian strain that was evolutionarily isolated from the typical wild waterfowl influenza gene pool for some time—one that, like the SARS coronavirus, emerged into circulation among humans from an as yet unknown animal host.

Future Investigations

OUR ANALYSES of five RNA segments from the 1918 virus have shed some light on its origin and strongly suggest that the pandemic virus was the common ancestor of both subsequent human and swine H1N1 lineages, rather than having emerged from swine. To date, analyzing the viral genes has offered no definitive clue to the exceptional virulence of the 1918 virus strain. But experiments with engineered viruses containing 1918 genes indicate that certain of the 1918 viral proteins could promote rapid virus replication and provoke an intensely destructive host immune response.

In future work, we hope that the 1918 pandemic virus strain can be placed in the context of influenza viruses that immediately preceded and followed it. The direct precursor of the pandemic virus, the first or spring wave virus strain, lacked the autumn wave's

avian sequences than do the 1957 H2 and 1968 H3 subtypes. Thus, we concluded, either the 1918 HA gene spent some length of time in an intermediate host where it accumulated many changes from the original avian sequence, or the gene came directly from an avian virus, but one that was markedly different from known avian H1 sequences.

To investigate the latter possibility that avian H1 genes might have changed substantially in the eight decades since the 1918 pandemic, we collaborated with scientists from the Smithsonian Institution's Museum of Natural History and Ohio State University. After examining many preserved birds from the era, our group isolated an avian subtype H1 influenza strain from a Brant goose

collected in 1917 and stored in ethanol in the Smithsonian's bird collections. As it turned out, the 1917 avian H1 sequence was closely related to modern avian North American H1 strains, suggesting that avian H1 sequences have changed little over the past 80 years. Extensive sequencing of additional wild bird H1 strains may yet identify a strain more similar to the 1918 HA, but it may be that no avian H1 will be found resembling the 1918 strain because, in fact, the HA did not reassort directly from a bird strain.

In that case, it must have had some intermediate host. Pigs are a widely suggested possibility because they are known to be susceptible to both human and avian viruses. Indeed, simultaneous outbreaks of influenza were seen in hu-

exceptional virulence and seemed to spread less easily. At present, we are seeking influenza RNA samples from victims of the spring wave to identify any genetic differences between the two strains that might help elucidate why the autumn wave was more severe. Similarly, finding pre-1918 human influenza RNA samples would clarify which gene segments in the 1918 virus were completely novel to humans. The unusual mortality among young people during the 1918 pandemic might be explained if the virus shared features with earlier circulating strains to which older people had some immunity. And finding samples of H1N1 from the 1920s and later would help us understand the 1918 virus's subsequent evolution into less virulent forms.

We must remember that the mechanisms by which pandemic flu strains originate are not yet fully understood. Because the 1957 and 1968 pandemic strains had avian-like HA proteins, it seems most likely that they originated in the direct reassortment of avian and human virus strains. The actual circumstances of those reassortment events have never been identified, however, so no one knows how long it took for the novel strains to develop into human pandemics.

The 1918 pandemic strain is even more puzzling, because its gene sequences are consistent neither with direct reassortment from a known avian strain nor with adaptation of an avian strain in swine. If the 1918 virus should prove to have acquired novel genes through a different mechanism than subsequent pandemic strains, this could have important public health implications. An alternative origin might even have contributed to the 1918 strain's exceptional virulence. Sequencing of many more avian influenza viruses and research into alternative intermediate hosts other than swine, such as poultry, wild birds or horses, may provide more clues to the 1918 pandemic's source. Until the origins of such strains are better understood, detection and prevention efforts may overlook the beginning of the next pandemic. $\boxed{\text{SA}}$

Persistence Pays Off

Visiting Alaska in the summer of 1949, Swedish medical student Johan Hultin met Lutheran missionaries in Fairbanks who told him of the 1918 flu pandemic's toll on Inuit villages. One, a tiny settlement on the Seward Peninsula called Teller Mission, was all but wiped out in November 1918. Overwhelmed missionaries had to call in the U.S. Army to help bury 72 victims' bodies in a mass grave, which they marked by two crosses.

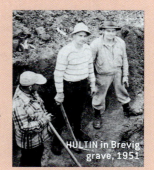
HULTIN in Brevig grave, 1951

Haunted by the story, Hultin (*right, center* and *below*) headed to the University of Iowa to begin his doctoral studies in microbiology. There he kept thinking about the 1918 pandemic and wondering if the deadly virus that caused it could be retrieved for study from bodies that may have been preserved by the Alaskan permafrost. In the summer of 1951, Hultin convinced two Iowa faculty, a virologist and a pathologist, to visit the village, then called Brevig Mission. With permission from tribal elders, the scientists excavated the grave and obtained tissue specimens from what remained of several victims' lungs.

Back in Iowa, the team tried and tried to grow live virus from the specimens but never could. In retrospect, that was perhaps just as well since biological containment equipment for dangerous pathogens did not exist at the time.

Hultin's disappointment led him to abandon his Ph.D. and become a pathologist instead. Retired and living in San Francisco in 1997, Hultin read our group's first published description of the 1918 genes we retrieved from autopsy specimens, and it rekindled his hope of finding the entire 1918 virus. He wrote to me, eager to try to procure new lung specimens from Brevig Mission for us to work with. He offered to leave immediately for Alaska, and I agreed.

At the same time, Hultin tracked down his 1951 expedition mates to ask if they had kept any of the original Brevig specimens. We reasoned that those tissue samples obtained just 33 years after the pandemic and then preserved might be in better condition than specimens taken later. As it turned out, one of Hultin's colleagues had kept the material in storage for years but finally deemed it useless and threw it out. He had disposed of the last specimens just the year before, in 1996.

Fortunately, Hultin once again got permission from the Brevig Mission Council to excavate the 1918 grave in August 1997. And this time he found the body of a young woman who had been obese in life. Hultin said later that he knew instantly her tissue samples would contain the 1918 virus—together with the cold temperature, her thick layer of fat had almost perfectly preserved her lungs. He was right, and her tissue provided us with the entire genome of the 1918 pandemic virus.   —*J.K.T.*

HULTIN in Brevig grave, 1997

MORE TO EXPLORE

Devil's Flu: The World's Deadliest Influenza Epidemic and the Scientific Hunt for the Virus That Caused It. Pete Davies. Henry Holt and Co., 2000.

America's Forgotten Pandemic: The Influenza of 1918. Second edition. Alfred W. Crosby. Cambridge University Press, 2003.

The Origin of the 1918 Pandemic Influenza Virus: A Continuing Enigma. Ann H. Reid and Jeffery K. Taubenberger in *Journal of General Virology*, Vol. 84, Part 9, pages 2285–2292; September 2003.

Global Host Immune Response: Pathogenesis and Transcriptional Profiling of Type A Influenza Viruses Expressing the Hemagglutinin and Neuraminidase Genes from the 1918 Pandemic Virus. J. C. Kash, C. F. Basler, A. Garcia-Sastre, V. Carter, R. Billharz, D. E. Swayne, R. M. Przygodzki, J. K. Taubenberger, M. G. Katze and T. M. Tumpey in *Journal of Virology*, Vol. 78, No. 17, pages 9499–9511; September 2004.

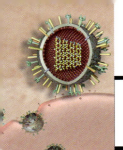

Article Review: CELL BIOLOGY

Capturing a Killer Flu Virus by Jeffery K. Taubenberger, Ann H. Reid and Thomas G. Fanning

TESTING YOUR COMPREHENSION

1. During the Spanish Flu epidemic (1918-1919), it is estimated that _____% of the world's population was infected with this strain of influenza virus. What percentage of infected individuals died as a result of infection with this strain of influenza?

    ANSWER: *The Spanish Flu epidemic is estimated to have infected ~30% of the world's population, killing ~5% of the infected individuals (total deaths ~20 to 50 million people).*

2. In the U.S. today, what is the typical percentage of people that are annually infected with an influenza virus? What percentage of infected individuals die as a result of their influenza infection?

    ANSWER: *In a typical influenza season, between 5% and 20% of the U.S. population will be infected with the virus. The lethality of the virus varies with each strain, but is usually ~0.1% or 1 death out of 1000 infected.*

3. What role does the hemagglutinin (HA) protein play in influenza A viral infections? How many isoforms of this protein are known?

    ANSWER: *Hemagglutinin proteins bind to cell surface receptors on cells infected by the virus. Specifically, HA proteins bind to distinct forms of N-Acetyl Neuraminic acid (NAN), also known as sialic acid, which are attached to the outer cell membrane of host cells. Different forms of the HA protein confer varied specificity for different forms of NAN, and have a critical impact upon the species infected. There are at least 15 isoforms of HA in the different strains of influenza A.*

4. What role does neuraminidase (NA) protein play in influenza A viral infections? How many isoforms of this protein are known?

    ANSWER: *The neuraminidase protein is mushroom shaped, extending outward from the membrane of each influenza virus. The neuraminidase protein is an active enzyme that cleaves cell surface–associated sialic acid (NAN) from host and viral carbohydrates during the process of viral release from an infected cell. Inhibition of this enzymatic activity is the mode of action for some antiviral drugs. There are at least 9 isoforms of NA in the different strains of influenza A.*

CLASS ACTIVITIES AND DISCUSSION

1. If you were trying to revive a virus from frozen tissue and grow it in your laboratory, what procedures might be involved? How would you determine if the virus was indeed growing in your culturing system? What safety measures would have to be taken if the virus was a risk to the health of laboratory workers?

2. Influenza virus has continued to be a major health threat to human populations despite a worldwide annual distribution of influenza vaccine, and large numbers of humans who have been exposed to influenza virus and have immunity to that strain of the virus. How has influenza continued to pose a serious threat to human health? As a class, identify the key viral characteristics that facilitate this phenomenon.

3. The reconstruction of the 1918 influenza virus was a major success for scientists hoping to understand the basis of its severe lethality. However, this virus does pose a health threat to human populations. In addition, publishing the genome sequence of a killer virus may be helpful to those who would abuse science to develop biological weapons. Do you agree or disagree with the scientists who feel that the reconstruction of the 1918 virus is justified by the prospective benefits of subsequent research?

Phillip Sharp (Massachusetts Institute of Technology) wrote an editorial to Science magazine discussing the risks of this work, but supporting the decision of the scientists to publish their work. Sharp PA (2005), 1918 flu and responsible science, Science 310:17. Editorial comments.

Article Review: GENETICS

Capturing a Killer Flu Virus *by Jeffery K. Taubenberger, Ann H. Reid and Thomas G. Fanning*

TESTING YOUR COMPREHENSION

1. The genomes of viruses vary widely with regard to composition (DNA/RNA), size, and the overall number of genes. How many genes are contained within the influenza genome? A human cell has ~30,000 genes — how is a virus able to infect a cell and replicate itself with so few genes?

   ANSWER: *The influenza genome contains only 11 genes carried on 8 separate segments of viral RNA. The encoded viral proteins mediate recognition and entry into host cells, replication of the viral genome, and exit from the infected cell. Viruses can use the host cell to manufacture viral components, thus reducing the number of genes required for the viral life cycle.*

2. Pigs can be simultaneously infected with both the avian influenza virus and human influenza viruses. Why does this phenomenon put human health at risk?

   ANSWER: *Pigs serve as a site for "viral mixing" where segments of the viral genome can be exchanged during infection with multiple influenza viruses. This can generate new forms of the virus, with new characteristics that may enable a more rapid spread of the virus or may make the virus more lethal.*

3. Influenza viruses have no proofreading mechanism to ensure proper replication of the viral genome during an infection. What benefit might this have for the long-term survival of the virus?

   ANSWER: *The high mutation rate in viruses ensure rapid changes in the virus structure and nature. This can enable evasion of the immune system as antigenic sites on the surface of the virus are continually changing.*

CLASS ACTIVITIES AND DISCUSSION

1. Of the 8 segments of the influenza A RNA genome, which segment encodes the neuraminidase gene? Go online, and use the National Center for Biotechnology Information (NCBI) website to examine the influenza A genome. Hint: use the GenBank component of NCBI to search for information or try http://www.ncbi.nlm.nih.gov/genomes/VIRUSES/viruses.html as a starting point.

2. You have recently isolated and sequenced a fragment of viral RNA from infected cells. You make a DNA copy of the fragment, the sequence of which is as follows: 5' - agcaaaagca ggggttatac catagacaac caaaagcata acaatggcca tcatttatct. Use GenBank and the BLAST program to search for known sequences that correspond to your isolated fragment. Is this an influenza virus sequence? What gene does this genome fragment encode?

3. In 2005, a team of scientists (including the authors of this article), were able to reassemble the 8 segments of 1918 influenza virus using reverse genetics and synthesized RNA (non-coding sequences were taken from a related influenza virus). Active viral particles were then made, essentially restoring the 1918 strain of influenza. If you were working in the laboratory that now had this reconstructed virus, what experiments might you perform? Why would these experiments be valuable? The article below describes some of the first experiments performed with this reconstructed virus.

   Tumpey TM, et al. (2005), Characterization of the reconstructed 1918 Spanish influenza pandemic virus, *Science* 310:77-80.

Web Resources:

Department of Health and Human Services – http://hhs.gov/flu/

Centers for Disease Control and Prevention – http://www.cdc.gov/flu/

Reuters website about influenza – http://www.influenza.com/

EXPLAINING
PANDEMIC
FLU

DO NOT
ENTER
BIOSECURITY
IN EFFECT

Tamiflu 75 mg

Tamiflu 75 mg

Tamiflu

FLIR
최고 33.7°
수동   36.5

4:08:30 p e=0.88 Trefl=25

One day a highly contagious and lethal strain of influenza will sweep across all humanity, claiming millions of lives. It may arrive in months or not for years—but the next pandemic is inevitable. **Are we ready?**

Preparing for a Pandemic

By W. Wayt Gibbs and Christine Soares

When the levees collapsed in New Orleans, the faith of Americans in their government's ability to protect them against natural disasters crumbled as well. Michael Chertoff, the secretary of homeland security who led the federal response, called Hurricane Katrina and the flood it spawned an "ultracatastrophe" that "exceeded the foresight of the planners."

But in truth the failure was not a lack of foresight. Federal, state and local authorities had a plan for how governments would respond if a hurricane were to hit New Orleans with 120-mile-per-hour winds, raise a storm surge that overwhelmed levees and water pumps, and strand thousands inside the flooded city. Last year they even practiced it. Yet when Katrina struck, the execution of that plan was abysmal.

The lethargic, poorly coordinated and undersized response raises concerns about how nations would cope with a much larger and more lethal kind of natural disaster that scientists warn will occur, possibly soon: a pandemic of influenza. The threat of a flu pandemic is more ominous, and its parallels to Katrina more apt, than it might first seem. The routine seasonal upsurges of flu and of hurricanes engender a familiarity that easily leads to complacency and inadequate preparations for the "big one" that experts admonish is sure to come.

The most fundamental thing to understand about serious pandemic influenza is that, except at a molecular level, the disease bears little resemblance to the flu that we all get at some time. An influenza pandemic, by definition, occurs only when the influenza virus mutates into something dangerously unfamiliar to our immune systems and yet is able to jump from person to person through a sneeze, cough or touch.

Flu pandemics emerge unpredictably every generation or so, with the last three striking in 1918, 1957 and 1968. They get their start when one of the many influenza strains that constantly circulate in wild and domestic birds evolves into a form that infects us as well. That virus then adapts further or exchanges genes with a flu strain native to humans to produce a novel germ that is highly contagious among people.

Some pandemics are mild. But some are fierce. If the virus replicates much faster than the immune system learns to defend against it, it will cause severe and sometimes fatal illness, resulting in a pestilence that could easily claim more lives in a single year than AIDS has in 25. Epidemiologists have warned that the next pandemic could sicken one in every three people on the planet, hospitalize many of those and kill tens to hundreds of millions. The disease would spare no nation, race or income group. There would be no certain way to avoid infection.

Scientists cannot predict which influenza strain will cause a pandemic or when the next one will break out. They can warn only that another is bound to come and that the conditions now seem ripe, with a fierce strain of avian flu killing people in Asia and infecting birds in a rapid westward lunge toward Europe. That strain, influenza A (H5N1) does not yet pass readily from one person to another. But the virus is evolving, and some of the affected avian species have now begun their winter migrations.

As a sense of urgency grows, governments and health experts are working to bolster four substantial lines of defense against a pandemic: surveillance, vaccines, containment measures and medical treatments. The U.S. plans to release by October a pandemic preparedness plan that surveys the strength of each of these barricades. Some failures are inevitable, but the more robust those preparations are, the less humanity will suffer. The experience of Katrina forces a question: Will authorities be able to keep to their plans even when a large fraction of their own workforce is downed by the flu?

EVOLUTION OF AN EPIDEMIC

H5N1 strain of avian flu sickens 18 people and kills 6 in Hong Kong

| 1918 | 1957 | 1968 | 1997 | 1999 |

Pandemic (H1N1 strain) kills 40 million worldwide

Pandemic (H2N2 strain) kills 1–4 million worldwide

Pandemic (H3N strain) kills 1 million worldwide

Surveillance: What Is Influenza Up to Now?

Our first defense against a new flu is the ability to see it coming. Three international agencies are coordinating the global effort to track H5N1 and other strains of influenza. The World Health Organization (WHO), with 110 influenza centers in 83 countries, monitors human cases. The World Organization for Animal Health (OIE, formerly the Office International des Épizooties) and the Food and Agriculture Organization (FAO) collect reports on outbreaks in birds and other animals. But even the managers of these surveillance nets acknowledge that they are still too porous and too slow.

Speed is of the essence when dealing with a fast-acting airborne virus such as influenza. Authorities probably have no realistic chance of halting a nascent pandemic unless they can contain it within 30 days [see "Rapid Response," on page 50]. The clock begins ticking the moment that the first victim of a pandemic-capable strain becomes contagious.

The only way to catch that emergence in time is to monitor constantly the spread of each outbreak and the evolution of the virus's abilities. The WHO assesses both those factors to determine where the world is in the pandemic cycle, which a new guide issued in April divides into six phases.

The self-limiting outbreaks of human H5N1 influenza seen so far bumped the alert level up to phase three, two steps removed from outright pandemic (phase six). Virologists try to obtain samples from every new H5N1 patient to scout for signs that the avian virus is adapting to infect humans more efficiently. It evolves in two ways: gradually through random mutation, and more rapidly as different strains of influenza swap genes inside a single animal or person [see box on opposite page].

The U.S. has a sophisticated flu surveillance system that funnels information on hospital visits for influenzalike illness, deaths from respiratory illness and influenza strains seen in public health laboratories to the Centers for Disease Control and Prevention in Atlanta. "But the system is not fast enough to take the isolation or quarantine action needed to manage avian flu," said Julie L. Gerberding, the CDC director, at a February conference. "So we have been broadening our networks of clinicians and veterinarians."

In several dozen cases where travelers to the U.S. from H5N1-affected Asian countries developed severe flulike symptoms, samples were rushed to the CDC, says Alexander Klimov of the CDC's influenza branch. "Within 40 hours of hospitalization we can say whether the patient has H5N1. Within another six hours we can analyze the genetic sequence of the hemagglutinin gene" to estimate the infectiousness of the strain. (The virus uses hemagglutinin to pry its way into cells.) A two-day test then reveals resistance to antiviral drugs, he says.

The next pandemic could break out anywhere, including in the U.S. But experts think it is most likely to appear first in Asia, as do most influenza strains that cause routine annual epidemics. Aquatic birds such as ducks and geese are the

Overview/*The Plan to Fight a New Flu*

- Scientists warn that a global epidemic caused by some newly evolved strain of influenza is inevitable and poses an enormous threat to public health.
- The pandemic could occur soon or not for years. H5N1 bird flu has killed more than 60 people in Asia, raising alarms. Even if that outbreak wanes, however, a global surveillance network must remain alert for other threatening strains.
- Flu shots matched to the new virus will arrive too late to prevent or slow the early stages of a pandemic, but rapid response with antiviral drugs might contain an emerging flu strain at its source temporarily, buying time for international preparations.
- Severity of disease will depend on the pandemic strain. In many places, drug supplies and other health resources will be overwhelmed.

9N2 infects children in ong Kong	Human H5N1 cases confirmed in Vietnam and Thailand	U.S. orders 2 million doses of H5N1 vaccine	6,000 wild birds die from H5N1 flu at a lake in central China	3 members of a suburban family in Indonesia die of H5N1	Vietnam begins immunization of 20 million fowl against H5N1	Since 2003, H5N1 has infected birds in 13 nations; people in 4		

3 / **2004: Jan** | **Sept** / **2005: April** | **June** | **July** | **August** | **September**

s H5N1 spreads to fowl in Asian nations, H7N7 infects ,000 people in the Netherlands	President Bush authorizes quarantine of people exposed to pandemic flu	Russia culls poultry as H5N1 flu spreads to Siberia	H5N1 flu found in geese flocks in Kazakhstan	Geese and swans found dead of H5N1 in Mongolia	Epidemic reaches fowl in the Ural Mountains in Russia

natural hosts for influenza, and in Asia many villagers reside cheek by bill with such animals. Surveillance in the region is still spotty, however, despite a slow trickle of assistance from the WHO, the CDC and other organizations.

A recent H5N1 outbreak in Indonesia illustrates both the problems and the progress. In a relatively wealthy suburb of Jakarta, the eight-year-old daughter of a government auditor fell ill in late June. A doctor gave her antibiotics, but her fever worsened, and she was hospitalized on June 28. A week later her father and one-year-old sister were also admitted to the hospital with fever and cough. The infant died on July 9, the father on July 12.

The next day an astute doctor alerted health authorities and sent blood and tissue samples to a U.S. Navy medical research unit in Jakarta. On July 14 the girl died; an internal report shows that on this same day Indonesian technicians in the naval laboratory determined that two of the three family members had H5N1 influenza. The government did not acknowledge this fact until July 22, however, after a WHO lab in Hong Kong definitively isolated the virus.

The health department then readied hospital wards for more flu patients, and I Nyoman Kandun, head of disease control for Indonesia, asked WHO staff to help investigate the outbreak. Had this been the onset of a pandemic, the 30-day containment window would by that time have closed. Kandun called off the investigation two weeks later. "We could not find a clue as to where these people got the infection," he says.

Local custom prohibited autopsies on the three victims. Klaus Stöhr of the WHO Global Influenza Program has complained that the near absence of autopsies on human H5N1 cases leaves many questions unanswered. Which organs does H5N1 infect? Which does it damage most? How strongly does the immune system respond?

Virologists worry as well that they have too little information about the role of migratory birds in transmitting the disease across borders. In July domestic fowl infected with H5N1 began turning up in Siberia, then Kazakhstan, then Russia. How the birds caught the disease remains a mystery.

Frustrated with the many unanswered questions, Stöhr and other flu scientists have urged the creation of a global task force to supervise pandemic preparations. The OIE in August appealed for more money to support surveillance programs it is setting up with the FAO and the WHO.

"We clearly need to improve our ability to detect the virus," says Bruce G. Gellin, who coordinates U.S. pandemic planning as head of the National Vaccine Program Office at the U.S. Department of Health and Human Services (HHS). "We need to invest in these countries to help them, because doing so helps everybody."

HOW A PANDEMIC STRAIN EMERGES

Avian strains of influenza A, such as H5N1, can evolve via two paths into pandemic-capable virus (able to bind readily to sialic acid on human cells). Genetic mutations and natural selection can render the virus more efficient at entering human cells (*pink path*). Alternatively (*yellow path*), two strains of influenza may infect the same cell (*a*) and release viral RNA, which replicates inside the cell nucleus (*b*). RNA from the two strains can then mix to create a set of "reassorted" genes (*c*) that give rise to a novel and highly contagious pandemic strain.

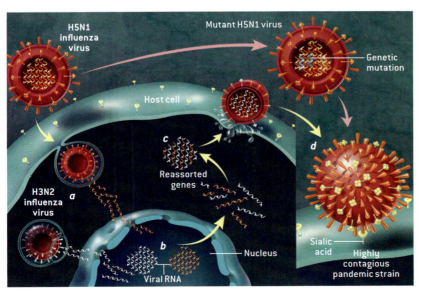

Vaccines: Who Will Get Them—and How Quickly?

Pandemics of smallpox and polio once ravaged humanity, but widespread immunization drove those diseases to the brink of extinction. Unfortunately, that strategy will not work against influenza—at least not without a major advance in vaccine technology.

Indeed, if an influenza pandemic arrives soon, vaccines against the emergent strain will be agonizingly slow to arrive and frustratingly short in supply. Biology, economics and complacency all contribute to the problem.

Many influenza strains circulate at once, and each is constantly evolving. "The better the match between the vaccine and the disease virus, the better the immune system can defend against the virus," Gellin explains. So every year manufacturers fashion a new vaccine against the three most threatening strains. Biologists first isolate the virus and then modify it using a process called reverse genetics to make a seed virus. In vaccine factories, robots inject the seed virus into fertilized eggs laid by hens bred under hygienic conditions. The pathogen replicates wildly inside the eggs.

Vaccine for flu shots is made by chemically dissecting the virus and extracting the key proteins, called antigens, that stimulate the human immune system to make the appropriate antibodies. A different kind of vaccine, one inhaled rather than injected, incorporates live virus that has been damaged enough that it can infect but not sicken. The process requires six months to transform viral isolates into initial vials of vaccine.

Because people will have had no prior exposure to a pandemic strain of influenza, everyone will need two doses: a primer and then a booster about four weeks later. So even those first in line for vaccines are unlikely to develop immunity until at least seven or eight months following the start of a pandemic.

And there will undoubtedly be a line. Total worldwide production of flu vaccine amounts to roughly 300 million doses a year. Most of that is made in Europe; only two plants operate in the U.S. Last winter, when contamination shut down a Chiron facility in Britain, Sanofi Pasteur and MedImmune pulled out all stops on their American lines—and produced 61 million doses. The CDC recommends annual flu immunization for high-risk groups that in the U.S. include some 185 million people.

Sanofi now runs its plant at full bore 365 days a year. In July it broke ground for a new facility in Pennsylvania that will double its output—in 2009. Even in the face of an emergency, "it would be very hard to compress that timeline," says James T. Matthews, who sits on Sanofi's pandemic-planning working group. He says it would not be feasible to convert factories for other kinds of vaccines over to make flu shots.

Pascale Wortley of the CDC's National Immunization Program raises another concern. Pandemics typically overlap with the normal flu season, she notes, and flu vaccine plants can make only one strain at a time. Sanofi spokesman Len Lavenda agrees that "we could face a Sophie's choice: whether to stop producing the annual vaccine in order to start producing the pandemic vaccine."

MedImmune aims to scale up production of its inhalable vaccine from about two million doses a year to 40 million doses by 2007. But Gellin cautions that it might be too risky to distribute live vaccine derived from a pandemic strain. There is a small chance, he says, that the virus in the vaccine could exchange genes with a "normal" flu virus in a person and generate an even more dangerous strain of influenza.

Because delays and shortages in producing vaccine against a pandemic are unavoidable, one of the most important functions of national pandemic plans is to push political leaders to decide in advance which groups will be the first to receive vaccine and how the government will enforce its rationing. The U.S. national vaccine advisory committee recommended in July that the first shots to roll off the lines should go to key government leaders, medical caregivers, workers in flu vaccine and drug factories, pregnant women, and those infants, elderly and ill people who are already in the high-priority group for an-

EGG-BASED PROCESS for making flu vaccine imposes bottlenecks that will delay the release of a pandemic vaccine by six months or more. Supplies will fall far short of the demand.

nual flu shots. That top tier includes about 46 million Americans.

Among CDC planners, Wortley says, "there is a strong feeling that we ought to say beforehand that the government will purchase some amount of vaccine to guarantee equitable distribution." Australia, Britain, France and other European governments are working out advance contracts with vaccine producers to do just that. The U.S., so far, has not.

In principle, governments could work around these supply difficulties by stockpiling vaccine. They would have to continually update their stocks as new strains of influenza threatened to go global; even doing so, the reserves would probably always be a step or two behind the disease. Nevertheless, Wortley says, "it makes sense to have H5N1 vaccine on hand, because even if it is not an exact match, it probably would afford some amount of protection" if the H5N1 strain evolved to cause a pandemic.

To that end, the U.S. National Institute of Allergy and Infectious Diseases (NIAID) last year distributed an H5N1 seed virus created from a victim in Vietnam by scientists at St. Jude Children's Research Hospital in Memphis. The HHS then placed an order with Sanofi for two million doses of vaccine against that strain. Human trials began in March, and "the preliminary results from the clinical trial indicate that the vaccine would be protective," says NIAID director Anthony S. Fauci. "HHS Secretary Michael Leavitt is trying to negotiate to get up to 20 million doses," he adds. (Leavitt announced in September that HHS had increased its H5N1 vaccine order by $100 million.) According to Gellin, current vaccine producers could contribute at most 15 million to 20 million doses a year to the U.S. stockpile.

Those numbers are probably over-optimistic, however. The trial tested four different concentrations of antigen. A typical annual flu shot has 45 micrograms of protein and covers three strains of influenza. Officials had expected that 30 micrograms of H5N1 antigen—two shots, with 15 micrograms in each—would be enough to induce immunity. But the preliminary trial results suggest

NEW VACCINE TECHNOLOGIES

Researchers in industry and academia are testing new immunization methods that would stretch the limited supply of vaccines to cover more people. They are also developing technologies that could allow vaccine production to increase rapidly in an emergency.

Technology	Benefits	Readiness	Companies
Intradermal injectors	Delivering flu vaccine into the skin rather than muscle might cut the required dose per shot by a factor of five	Clinical trials show promise, but few nurses and doctors are trained in the procedure	Iomai, GlaxoSmithKline
Adjuvants	Chemical additives called adjuvants can increase the immune response, so that less protein is needed per shot	One such vaccine is licensed in Europe. Others are in active development	Iomai, Chiron, GlaxoSmithKline
Cell-cultured vaccines	Growing influenza virus for vaccine in cell-filled bioreactors, rather than in eggs, would enable faster increases in production if a flu pandemic broke out	Chiron is conducting a large-scale trial in Europe. Sanofi Pasteur and Crucell are developing a process for the U.S.	Chiron, Baxter, Sanofi Pasteur, Crucell, Protein Sciences
DNA vaccines	Gold particles coated with viral DNA could be injected into the skin with a jet of air. Production of DNA vaccines against a new strain could begin in weeks, rather than months. Stockpiles would last years without refrigeration	No DNA vaccine has yet been proved effective in humans. PowderMed expects results from a small-scale trial of an H5N1 DNA vaccine in late 2006	PowderMed, Vical
All-strain vaccines	A vaccine that raises immunity against a viral protein that rarely mutates might thwart every strain of influenza. Stockpiles could then reliably defend against a pandemic	Acambis began developing a vaccine against the M2e antigen this past summer	Acambis

that 180 micrograms of antigen are needed to immunize one person.

An order for 20 million conventional doses may thus actually yield only enough H5N1 vaccine for about 3.3 million people. The true number could be even lower, because H5 strains grow poorly in eggs, so each batch yields less of the active antigen than usual. This grim picture may brighten, however, when NIAID analyzes the final results from the trial. It may also be possible to extend vaccine supplies with the use of adjuvants (substances added to vaccines to increase the immune response they induce) or new immunization approaches, such as injecting the vaccine into the skin rather than into muscle.

Caching large amounts of prepandemic vaccine, though not impossible, is clearly a challenge. Vaccines expire after a few years. At current production rates, a stockpile would never grow to the 228

million doses needed to cover the three highest priority groups, let alone to the roughly 600 million doses that would be needed to vaccinate everyone in the U.S. Other nations face similar limitations.

The primary reason that capacity is so tight, Matthews explains, is that vaccine makers aim only to meet the demand for annual immunizations when making business decisions. "We really don't see the pandemic itself as a market opportunity," he says.

To raise manufacturers' interest, "we need to offer a number of incentives, ranging from liability insurance to better profit margins to guaranteed purchases," Fauci acknowledges. Long-term solutions, Gellin predicts, may come from new technologies that allow vaccines to be made more efficiently, to be scaled up more rapidly, to be effective at much lower doses and perhaps to work equally well on all strains of influenza.

Rapid Response: Could a Pandemic Be Stopped?

As recently as 1999, WHO had a simple definition for when a flu pandemic began: with confirmation that a new virus was spreading between people in at least one country. Thereafter, stopping the flu's lightning-fast expansion was unthinkable—or so it then seemed. But because of recent advances in the state of disease surveillance and antiviral drugs, the latest version of WHO's guidelines recognizes a period on the cusp of the pandemic when a flu virus ready to burst on the world might instead be intercepted and restrained, if not stamped out.

Computer models and common sense indicate that a containment effort would have to be exceptionally swift and efficient. Flu moves with extraordinary speed because it has such a short incubation period—just two days after infection by the virus, a person may start showing symptoms and shedding virus particles that can infect others. Some people may become infectious a day before their symptoms appear. In contrast, people infected by the SARS coronavirus that emerged from China in 2003 took as long as 10 days to become infectious, giving health workers ample time to trace and isolate their contacts before they, too, could spread the disease.

Contact tracing and isolation alone could never contain flu, public health experts say. But computer-simulation results published in August showed when up to 30 million doses of antiviral drugs and a low-efficacy vaccine were added to the interventions a chance emerged to thwart a potential pandemic.

Conditions would have to be nearly ideal. Modeling a population of 85 million based on the demographics and geography of Thailand, Neil M. Ferguson of Imperial College London found that health workers would have at most 30 days from the start of person-to-person viral transmission to deploy antivirals as both treatment and preventives wherever outbreaks were detected.

But even after seeing the model results earlier this year, WHO officials expressed doubt that surveillance in parts of Asia is reliable enough to catch a budding epidemic in time. In practice, confirmation of some human H5N1 cases has taken more than 20 days, WHO flu chief Stöhr warned a gathering of experts in Washington, D.C., this past April. That leaves just a narrow window in which to deliver the drugs to remote areas and dispense them to as many as one million people.

Partial immunity in the population could buy more time, however, according to Ira M. Longini, Jr., of Emory University. He, too, modeled intervention with antivirals in a smaller community based on Thai demographic data, with outcomes similar to Ferguson's. But Longini added scenarios in which people had been vaccinated in advance. He assumed that an existing vaccine, such as the H5N1 prototype version some countries have already developed, would not perfectly match a new variant of the virus, so his model's vaccinees were only 30 percent less likely to be infected. Still, their reduced susceptibility made containing even a highly infectious flu strain possible in simulations. NIAID director Fauci has said that the U.S. and other nations with H5N1 vaccine are still considering whether to direct it toward prevention in the region where a human-adapted version of that virus is most likely to emerge—even if that means less would remain for their own citizens. "If we're smart, we would," Longini says.

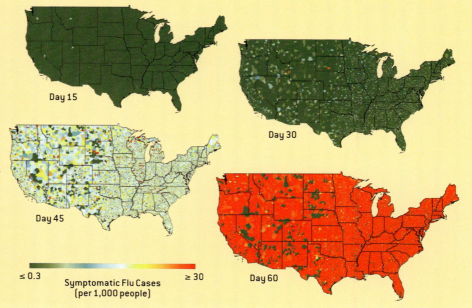

Pandemic Flu Hits the U.S.

A simulation created by researchers from Los Alamos National Laboratory and Emory University shows the first wave of a pandemic spreading rapidly with no vaccine or antiviral drugs employed to slow it down. Colors represent the number of symptomatic flu cases per 1,000 people (see scale). Starting with 40 infected people on the first day, nationwide cases peak around day 60, and the wave subsides after four months with 33 percent of the population having become sick. The scientists are also modeling potential interventions with drugs and vaccines to learn if travel restrictions, quarantines and other disruptive disease-control strategies could be avoided.

Day 15

Day 30

Day 45

Day 60

≤ 0.3   Symptomatic Flu Cases (per 1,000 people)   ≥ 30

COMMAND CENTER at the U.S. Department of Health and Human Services in Washington, D.C., would be used to track the spread of a flu pandemic. From this location, HHS would coordinate the activities of its divisions, including the CDC and the NIH, and share information with state and federal government agencies such as the Department of Homeland Security.

Based on patterns of past pandemics, experts expect that once a new strain breaks loose, it will circle the globe in two or three waves, each potentially lasting several months [see box on opposite page] but peaking in individual communities about five weeks after its arrival. The waves could be separated by as long as a season: if the first hit in springtime, the second might not begin until late summer or early fall. Because meaningful amounts of vaccine tailored to the pandemic strain will not emerge from factories for some six months, government planners are especially concerned with bracing for the first wave.

Once a pandemic goes global, responses will vary locally as individual countries with differing resources make choices based on political priorities as much as on science. Prophylactic use of antivirals is an option for a handful of countries able to afford drug stockpiles, though not a very practical one. No nation has enough of the drugs at present to protect a significant fraction of its population for months. Moreover, such prolonged use has never been tested and could cause unforeseen problems. For these reasons, the U.K. declared this past July that it would use its pandemic

stockpile primarily for treating patients rather than for protecting the uninfected. The U.S., Canada and several other countries are still working out their priorities for who will receive antivirals and when.

For most countries there will be no choice: what the WHO calls nonpharmaceutical interventions will have to be their primary defense. Although the effectiveness of such measures has not been extensively researched, the WHO gathered flu specialists in Geneva in March 2004 to try to determine which actions medical evidence does support. Screening incoming travelers for flu symptoms, for instance, "lacks proven health benefit," the group concluded, although they acknowledged that countries might do it anyway to promote public confidence. Similarly, they were skeptical that public fever screening, fever hotlines or fever clinics would do much to slow the spread of the disease.

The experts recommended surgical masks for flu patients and health workers exposed to those patients. For the healthy, hand washing offers more protection than wearing masks in public, because people can be exposed to the virus at home, at work and by touching contaminated surfaces—including the surface of a mask.

Traditional "social distancing" measures, such as banning public gatherings or shutting down mass transit, will have to be guided by what epidemiologists find once the pandemic is under way. If children are especially susceptible to the virus, for example—as was the case in 1957 and 1968—or if they are found to be an important source of community spread, then governments may consider closing schools.

Treatment: What Can Be Done for the Sick?

If two billion become sick, will 10 million die? Or 100 million? Public health specialists around the world are struggling to quantify the human toll of a future flu pandemic. Casualty estimates vary so widely because until it strikes, no one can be certain whether the next pandemic strain will be mild, like the 1968 virus that some flu researchers call a "wimp"; moderately severe, like the 1957 pandemic strain; or a stone-cold killer, like the "Great Influenza" of 1918.

For now, planners are going by rules of thumb: because no one would have immunity to a new strain, they expect 50 percent of the population to be infected

by the virus. Depending on its virulence, between one third and two thirds of those people will become sick, yielding a clinical attack rate of 15 to 35 percent of the whole population. Many governments are therefore trying to prepare for a middle-ground estimate that 25 percent of their entire nation will fall ill.

No government is ready now. In the U.S., where states have primary responsibility for their residents' health, the Trust for America's Health (TFAH) estimates that a "severe" pandemic virus sickening 25 percent of the population could translate into 4.7 million Americans needing hospitalization. The TFAH notes that the country currently

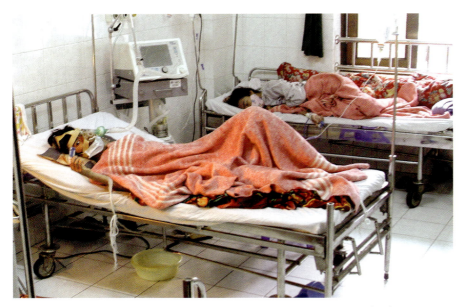

AVIAN FLU PATIENTS at a hospital in Hanoi, Vietnam, this past March were a man (*left*), 21, in critical condition, and his sister, 14. Many of the most severe illnesses and deaths from H5N1 infection have been among previously healthy young adults and children.

has fewer than one million staffed hospital beds.

For frontline health workers, a pandemic's severity will boil down to the sheer number of patients and the types of illness they are suffering. These, in turn, could depend on both inherent properties of the virus and susceptibility of various subpopulations to it, according to Maryland's pandemic planner, Jean Taylor. A so-called mild pandemic, for example, might resemble seasonal flu but with far larger numbers infected.

Ordinarily, those hardest hit by annual flu are people who have complications of chronic diseases, as well as the very young, the very old and others with weak immune systems. The greatest cause of seasonal flu-related deaths is pneumonia brought on by bacteria that invade after flu has depleted the body's defenses, not by the flu virus itself. Modeling a pandemic with similar qualities, Dutch national health agency researchers found that hospitalizations might be reduced by 31 percent merely by vaccinating the usual risk groups against bacterial pneumonia in advance.

In contrast, the 1918 pandemic strain was most lethal to otherwise healthy young adults in their 20s and 30s, in part because their immune systems were so hardy. Researchers studying that virus

have discovered that it suppresses early immune responses, such as the body's release of interferon, which normally primes cells to resist attack. At the same time, the virus provokes an extreme immune overreaction known as a cytokine storm, in which signaling molecules called cytokines summon a ferocious assault on the lungs by immune cells.

Doctors facing the same phenomenon in SARS patients tried to quell the storm by administering interferon and cytokine-suppressing corticosteroids. If the devastating cascade could not be stopped in time, one Hong Kong physician reported, the patients' lungs became increasingly inflamed and so choked with dead tissue that pressurized ventilation was needed to get enough oxygen to the bloodstream.

Nothing about the H5N1 virus in its current form offers reason to hope that it would produce a wimpy pandemic, according to Frederick G. Hayden, a University of Virginia virologist who is advising WHO on treating avian flu victims. "Unless this virus changes dramatically in pathogenicity," he asserts, "we will be confronted with a very lethal strain." Many H5N1 casualties have suffered acute pneumonia deep in the lower lungs caused by the virus itself, Hayden says, and in some cases blood

tests indicated unusual cytokine activity. But the virus is not always consistent. In some patients, it also seems to multiply in the gut, producing severe diarrhea. And it is believed to have infected the brains of two Vietnamese children who died of encephalitis without any respiratory symptoms.

Antiviral drugs that fight the virus directly are the optimal treatment, but many H5N1 patients have arrived on doctors' doorsteps too late for the drugs to do much good. The version of the strain that has infected most human victims is also resistant to an older class of antivirals called amantadines, possibly as a result of those drugs having been given to poultry in parts of Asia. Laboratory experiments indicate that H5N1 is still susceptible to a newer class of antivirals called neuraminidase inhibitors (NI) that includes two products, oseltamivir and zanamivir, currently on the market under the brand names Tamiflu and Relenza. The former comes in pill form; the latter is a powder delivered by inhaler. To be effective against seasonal flu strains, either drug must be taken within 48 hours of symptoms appearing.

The only formal test of the drugs against H5N1 infection, however, has been in mice. Robert G. Webster of St. Jude Children's Research Hospital reported in July that a mouse equivalent of the normal human dose of two Tamiflu pills a day eventually subdued the virus, but the mice required treatment for eight

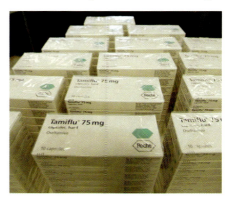

OSELTAMIVIR, sold as Tamiflu, is manufactured in a complex multistep process that takes nearly a year. Current stockpile orders will require several years to fill, and generic versions would be difficult to create for an emergency.

NEW FLU DRUGS

Today's flu antivirals disable specific proteins on the virus's surface—either M2 (drugs known as amantadines) or neuraminidase (zanamivir and oseltamivir). Some new drugs in development are improved neuraminidase inhibitors. Other novel approaches include blocking the virus's entry into host cells or hobbling its ability to function once inside.

Approach	Drugs	Benefits	Readiness
Inhibition of neuraminidase protein, which the virus uses to detach from one cell and infect another	Peramivir (BioCryst Pharmaceuticals); CS-8958 (Biota/Sankyo)	Neuraminidase inhibitors have fewer side effects and are less likely to provoke viral resistance than the older amantadines. CS-8958 is a long-acting formulation that clings inside lungs for up to a week	Peramivir reached lungs inefficiently in clinical trials of a pill form; trials of intravenous delivery may occur in 2006; initial safety trials are complete on CS-8958
Inhibition of viral attachment to cells	Fludase (NexBio)	Because it blocks the sialic acid receptor that flu viruses use to enter host cells, Fludase should be equally effective on all flu strains	Clinical trials are planned for 2006
Stimulation of RNA interference mechanism	GO0101 (Galenea); unnamed (Alnylam Pharmaceuticals)	Uses DNA to activate a built-in defense mechanism in cells, marking viral instructions for destruction. GO01498 demonstrated effective against avian H5 and H7 flu viruses in mice	Clinical trials are expected within 18 months
Antisense DNA to block viral genes	Neugene (AVI BioPharma)	Synthetic strands of DNA bind to viral RNA that instructs the host cell to make more virus copies. The strategy should be effective against most strains	Animal testing is scheduled for 2006

days rather than the usual five. The WHO is organizing studies of future H5N1 victims to determine the correct amount for people.

Even at the standard dosage, however, treating 25 percent of the U.S. population would require considerably more Tamiflu, or its equivalent, than the 22 million treatment courses the U.S. Department of Health and Human Services planned to stockpile as of September. An advisory committee has suggested a minimum U.S. stockpile of 40 million treatment courses (400 million pills). Ninety million courses would be enough for a third of the population, and 130 million would allow the drugs to also be used to protect health workers and other essential personnel, the committee concluded.

Hayden hopes that before a pandemic strikes, a third NI called peramivir may be approved for intravenous use in hospitalized flu patients. Long-acting NIs might one day be ideal for stockpiling because a single dose would suffice for treatment or offer a week's worth of prevention.

These additional drugs, like a variety of newer approaches to fighting flu

[see box above], all have to pass clinical testing before they can be counted on in a pandemic. Researchers would also like to study other treatments that directly modulate immune system responses in flu patients. Health workers will need every weapon they can get if the enemy they face is as deadly as H5N1.

Fatality rates in diagnosed H5N1 victims are running about 50 percent. Even if that fell to 5 percent as the virus traded virulence for transmissibility among people, Hayden warns, "it would still represent a death rate double [that of] 1918, and that's despite modern technologies like antibiotics and ventilators." Expressing the worry of most flu experts at this pivotal moment for public health, he cautions that "we're well behind the curve in terms of having plans in place

and having the interventions available."

Never before has the world been able to see a flu pandemic on the horizon or had so many possible tools to minimize its impact once it arrives. Some mysteries do remain as scientists watch the evolution of a potentially pandemic virus for the first time, but the past makes one thing certain: even if the dreaded H5N1 never morphs into a form that can spread easily between people, some other flu virus surely will. The stronger our defenses, the better we will weather the storm when it strikes. "We have only one enemy," CDC director Gerberding has said repeatedly, "and that is complacency." ⬛

W. Wayt Gibbs is senior writer. Christine Soares is a staff writer and editor.

MORE TO EXPLORE

The Great Influenza. Revised edition. John M. Barry. Penguin Books, 2005.

John R. LaMontagne Memorial Symposium on Pandemic Influenza Research: Meeting Proceedings. Institute of Medicine. National Academies Press, 2005.

WHO Global Influenza Preparedness Plan. WHO Department of Communicable Disease Surveillance and Response Global Influenza Program, 2005. www.who.int/csr/resources/publications/influenza/WHO_CDS_CSR_GIP_2005_5/en/index.html

Pandemic influenza Web site of the U.S. Department of Health and Human Services, National Vaccine Program Office: **www.hhs.gov/nvpo/pandemics/index.html**

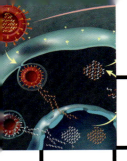

Article Review: CELL BIOLOGY

Preparing for a Pandemic   *by W. Wayt Gibbs and Christine Soares*

TESTING YOUR COMPREHENSION

1. Define the characteristics of the anticipated influenza pandemic. How is this different from the annual worldwide spread of influenza?

   ANSWER: *A pandemic is a rapid increase in the occurrence of a disease in a large and geographically dispersed population. The anticipated influenza pandemic would be a much more severe form of the disease; more lethal than the common influenza we currently experience, and something for which our immune systems have no prior immunity.*

2. Define the four key lines of defense that the government is working to strengthen over the coming years. What component of this plan requires the most time to develop?

   ANSWER: *The federal plan specifies funding to improve surveillance, vaccination, containment measures, and medical treatments. The development of a vaccination program for a population the size of the U.S. is a challenging task, and will require some time.*

3. The article describes a defined prioritized list of people who should be the first to be vaccinated from the available doses of H5N1 vaccine. The list was developed by the U.S. national vaccine advisory committeee. Define this prioritized list.

   ANSWER: *The list specifies that key government leaders, medical caregivers, workers in flu vaccine and drug factories, pregnant women, the elderly, and individuals who are already ill and have elevated susceptibility to contagions.*

CLASS ACTIVITIES AND DISCUSSION

1. The Department of Health and Human Services has posted the Federal Flu Plan announced by President Bush in October 2005. This plan will cost taxpayers 7.1 billion dollars, as the medical community in the U.S. prepares for the anticipated influenza pandemic. The document describes the establishment of a Strategic National Stockpile (SNS) of vaccine as rapidly as possible. Estimates anticipate 20 million doses within two or three years, and enough for the entire population by 2010. If this virus emerges as a national health crisis before 2010, how should the available vaccine be distributed?

   http://www.hhs.gov/pandemicflu/plan/

   Kaiser J (2005), Pandemic or Not, Experts Welcome the Bush Flu Plan, *Science* 310:952-3.

2. As a class, listen to the National Public Radio, Science Friday program that was initially broadcast on November 4, 2005. There are two sessions of the broadcast, a discussion of the federal flu plan and a discussion of vaccine production as it relates to this issue. Discuss the key comments, both in support and in criticism of the federal plan. Should the federal government do more to protect the public? If so, what should the government be doing?

   http://www.sciencefriday.com/pages/

3. A non-toxic, commercially available fluorescent material (Glo Germ™, http://www.glogerm.com/) offers a simple reagent for revealing how rapidly a virus or bacterium can spread from person to person. Glo Germ™ can be purchased as a powder, oil, or lotion that fluoresces brightly when exposed to a handheld UV light. Divide the class into small groups of students and develop a teaching module for an elementary school-age class. Use Glo-Germ™ to reveal the benefits of hand-washing, and general hygiene. Develop the module to have an active component where students participate in the demonstration, and a discussion component where students discuss/interpret the results with the instructor. Write this module up so anyone can use your instructions to teach it, and then present the module to your classmates.

Article Review: GENETICS

TESTING YOUR COMPREHENSION

1. With regard to pandemic influenza, what are the risks associated with a live, attenuated viral vaccine that is delivered by inhalation?

   ANSWER: *Though attenuated influenza vaccine can be quite effective for immunization, there is a perceived risk of recombination between the attenuated virus and other influenza that may be present in the patient. If the H5N1 strain were to recombine in this way, there is a chance that a more contagious and deadly virus could be generated.*

2. If a patient presents symptoms of a severe influenza that is suspected to be H5N1, what is the timeline to a confirmed diagnosis in the United States?

   ANSWER: *If all procedures and samples are handled efficiently, the process of confirming H5N1 can take as little as 2 days according to experts at the Centers for Disease Control. However, outside of the United States, cases have been slow to be diagnosed, sometimes taking up to a month before confirmation.*

CLASS ACTIVITIES AND DISCUSSION

1. In light of the difficulties associated with traditional vaccine production, researchers are evaluating several alternative strategies for producing an H5N1 vaccine for a large population. One strategy involves the use of a "DNA vaccine" where a portion of the viral genome is placed onto microscopic gold beads and shot into the skin of the vaccine recipient. The viral genes can be expressed in the cells that receive the gold beads, and those proteins can generate an immune response. The vaccination requires only a small amount of DNA. What benefits/limitations would you associate with DNA vaccines? Should the government turn from traditional, slower but successful methodologies for vaccine production and use its resources to develop a process like a DNA vaccine?

2. H5N1 has shown to be resistant to amantadine therapy, presumably because of the use of this drug to treat infected poultry. Explain the development of a drug-resistant strain of virus resulting from the use of that drug upon infected organisms. Does this parallel the development of antibiotic-resistant bacteria in populations where antibiotics are heavily prescribed?

3. Break the class up into 5 groups and assign each group a specific category of anti-influenza drug: 1) M2/amantadines, 2) neuraminidase inhibitors, 3) sialic acid receptor blockers (Fludase), 4) RNA interference (G00101), or 5) antisense DNA (Neugene). Each group will research their category of drug, assess the known and anticipated benefit, and define the mechanism of action for each drug. Groups will then present their results to the class. At the conclusion of the presentations, the class will discuss the 5 categories of treatment, and assess which category is most likely to be effective against an H5N1 pandemic.

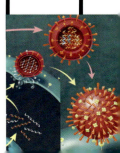

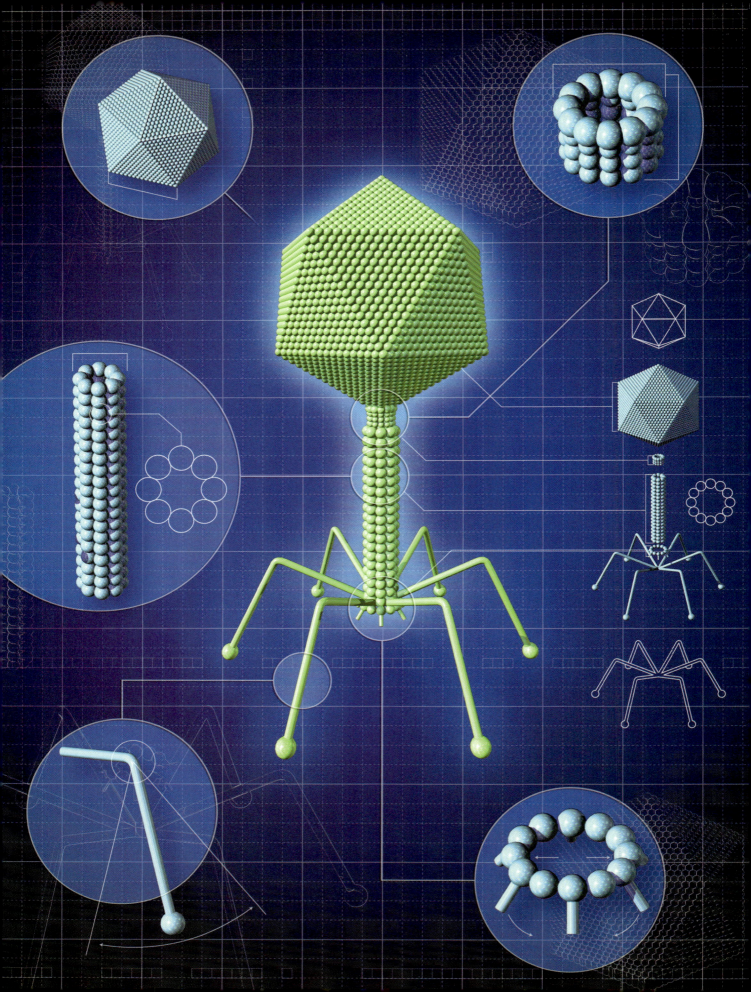

Biologists are crafting libraries of interchangeable DNA parts and assembling them inside microbes to create programmable, living machines

SYNTHETIC LIFE

By W. Wayt Gibbs

Evolution is a wellspring of creativity; 3.6 billion years of mutation and competition have endowed living things with an impressive range of useful skills. But there is still plenty of room for improvement. Certain microbes can digest the explosive and carcinogenic chemical TNT, for example—but wouldn't it be handy if they glowed as they did so, highlighting the location of buried land mines or contaminated soil? Wormwood shrubs generate a potent medicine against malaria but only in trace quantities that are expensive to extract. How many millions of lives could be saved if the compound, artemisinin, could instead be synthesized cheaply by vats of bacteria? And although many cancer researchers would trade their eyeteeth for a cell with a built-in, easy-to-read counter that ticks over reliably each time it divides, nature apparently has not deemed such a thing fit enough to survive in the wild.

It may seem a simple matter of genetic engineering to rewire cells to glow in the presence of a particular toxin, to manufacture an intricate drug, or to keep track of the cells' age. But creating such biological devices is far from easy. Biologists have been transplanting genes from one species to another for 30 years, yet genetic engineering is still more of a craft than a mature engineering discipline.

"Say I want to modify a plant so that it changes color in the presence of TNT," posits Drew Endy, a biologist at the Massachusetts Institute of Technology. "I can start tweaking genetic pathways in the plant to do that, and if I am lucky, then after a year or two I may get a 'device'— one system. But doing that once doesn't help me build a cell that swims around and eats plaque from artery walls. It doesn't help me grow a little microlens. Basically the current practice produces pieces of art."

Endy is one of a small but rapidly growing number of scientists who have set out in recent years to buttress the foundation of genetic engineering with what they call synthetic biology. They are designing and building living systems that behave in predictable ways, that use interchangeable parts, and in some cases that operate with an expanded genetic code, which allows them to do things that no natural organism can.

This nascent field has three major goals: One, learn about life by building it, rather than by tearing it apart. Two, make genetic engineering worthy of its

REDESIGNED VIRUSES will help biologists learn how to build reliable genetic machines. A group at the Massachusetts Institute of Technology has reorganized the genome of the T7 bacteriophage drawn here.

DREW ENDY (*pictured*) and others at M.I.T. have designed and built more than 140 "BioBricks" (*in vials*). Each is a piece of DNA that performs a well-characterized function and interacts well with other genetic parts.

name—a discipline that continuously improves by standardizing its previous creations and recombining them to make new and more sophisticated systems. And three, stretch the boundaries of life and of machines until the two overlap to yield truly programmable organisms. Already TNT-detecting and artemisinin-producing microbes seem within reach. The current prototypes are relatively primitive, but the vision is undeniably grand: think of it as Life, version 2.0.

A Light Blinks On

THE ROOTS OF SYNTHETIC BIOLOGY extend back 15 years to pioneering work by Steven A. Benner and Peter G. Schultz. In 1989 Benner led a team at ETH Zurich that created DNA containing two artificial genetic "letters" in addition to the four that appear in life as we know it. He and others have since invented several varieties of artificially enhanced DNA. So far no one has made genes from altered DNA that are functional—transcribed to RNA and then translated to protein form—within living cells. Just within the past year, however, Schultz's group at the Scripps Research Institute developed cells (containing normal DNA) that generate unnatural amino acids and string them together to make novel proteins [*see box on page 52*].

Overview/*Synthetic Biology*

- Molecular biology has been largely a reductive science that deduces the operation of living systems by breaking them apart.
- A growing number of synthetic biologists are taking a different approach: building machines from interchangeable DNA parts. The devices work inside living cells, from which they derive energy, raw materials, and the ability to move and reproduce.
- Synthetic biology has already produced microbes with a variety of unnatural talents. Some produce complex chemical ingredients for drugs; others make artificial amino acids, remove heavy metals from wastewater or perform simple binary logic.

Benner and other "old school" synthetic biologists see artificial genetics as a way to explore basic questions, such as how life got started on earth and what forms it may take elsewhere in the universe. Interesting as that is, the recent buzz growing around synthetic biology arises from its technological promise as a way to design and build machines that work inside cells. Two such devices, reported simultaneously in 2000, inspired much of the work that has happened since.

Both devices were constructed by inserting selected DNA sequences into *Escherichia coli*, a normally innocuous bacterium in the human gut. The two performed very different functions, however. Michael Elowitz and Stanislaus Leibler, then at Princeton University, assembled three interacting genes in a way that made the *E. coli* blink predictably, like microscopic Christmas tree lights [*see box on opposite page*]. Meanwhile James J. Collins, Charles R. Cantor and Timothy S. Gardner of Boston University made a genetic toggle switch. A negative feedback loop—two genes that interfere with each other—allows the toggle circuit to flip between two stable states. It effectively endows each modified bacterium with a rudimentary digital memory.

To engineering-minded biologists, these experiments were energizing but also frustrating. It had taken nearly a year to create the toggle switch and about twice that time to build the flashing microbes. And no one could see a way to connect the two devices to make, for example, blinking bacteria that could be switched on and off.

"We would like to be able to routinely assemble systems from pieces that are well described and well behaved," Endy remarks. "That way, if in the future someone asks me to make an organism that, say, counts to 3,000 and then turns left, I can grab the parts I need off the shelf, hook them together and predict how they will perform." Four years ago parts such as these were just a dream. Today they fill a box on Endy's desk.

Building with BioBricks

"THESE ARE GENETIC PARTS," Endy says as he holds out a container filled with more than 50 vials of clear, syrupy fluid. "Each of these vials contains copies of a distinct section of DNA that either performs some function on its own or can be

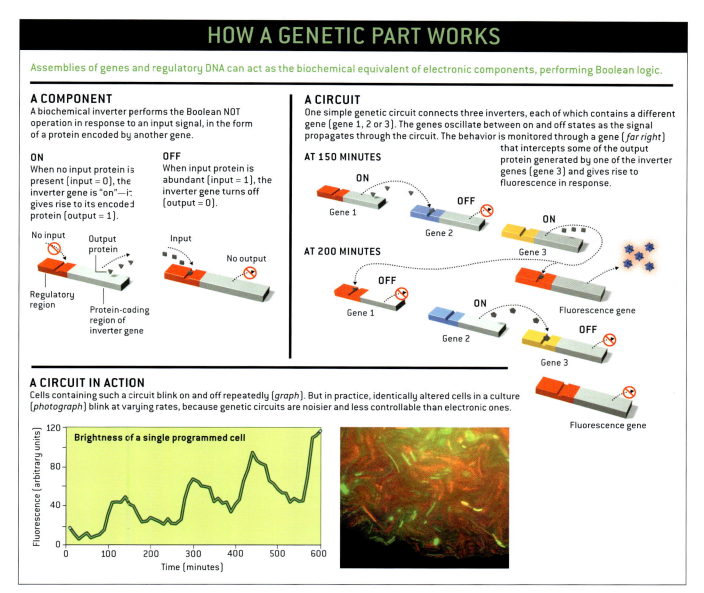

HOW A GENETIC PART WORKS

Assemblies of genes and regulatory DNA can act as the biochemical equivalent of electronic components, performing Boolean logic.

A COMPONENT

A biochemical inverter performs the Boolean NOT operation in response to an input signal, in the form of a protein encoded by another gene.

ON

When no input protein is present (input = 0), the inverter gene is "on"—it gives rise to its encoded protein (output = 1).

OFF

When input protein is abundant (input = 1), the inverter gene turns off (output = 0).

No input
Output protein
Input
No output
Regulatory region
Protein-coding region of inverter gene

A CIRCUIT

One simple genetic circuit connects three inverters, each of which contains a different gene (gene 1, 2 or 3). The genes oscillate between on and off states as the signal propagates through the circuit. The behavior is monitored through a gene (*far right*) that intercepts some of the output protein generated by one of the inverter genes (gene 3) and gives rise to fluorescence in response.

AT 150 MINUTES

ON
Gene 1
OFF
Gene 2
ON
Gene 3

AT 200 MINUTES

OFF
Gene 1
ON
Gene 2
OFF
Gene 3

Fluorescence gene

Fluorescence gene

A CIRCUIT IN ACTION

Cells containing such a circuit blink on and off repeatedly (*graph*). But in practice, identically altered cells in a culture (*photograph*) blink at varying rates, because genetic circuits are noisier and less controllable than electronic ones.

Brightness of a single programmed cell

Fluorescence (arbitrary units) — 0, 40, 80, 120
Time (minutes) — 0, 100, 200, 300, 400, 500, 600

---

used by a cell to make a protein that does something useful. What is important here is that each genetic part has been carefully designed to interact well with other parts, on two levels." At a mechanical level, individual BioBricks (as the M.I.T. group calls the parts) can be fabricated and stored separately, then later stitched together to form larger bits of DNA. And on a functional level, each part sends and receives standard biochemical signals. So a scientist can change the behavior of an assembly just by substituting a different part at a given spot.

"Interchangeable components are something we take for granted in other kinds of engineering," Endy notes, but genetic engineering is only beginning to draw on the power of the concept. One advantage it offers is abstraction. Just as electrical engineers need not know what is inside a capacitor before they use it in a circuit, biological engineers would like to be able to use a genetic toggle switch while remaining blissfully ignorant of the binding coefficients and biochemical makeup of the promoters, repressors, activators, inducers and other genetic el-

ements that make the switch work. One of the vials in Endy's box, for example, contains an inverter BioBrick (also called a NOT operator). When its input signal is high, its output signal is low, and vice versa. Another BioBrick performs a Boolean AND function, emitting an output signal only when it receives high levels of both its inputs. Because the two parts work with compatible signals, connecting them creates a NAND (NOT AND) operator. Virtually any binary computation can be performed with enough NAND operators.

Beyond abstraction, standardized parts offer another powerful advantage: the ability to design a functional genetic system without knowing exactly how to make it. Early last year a class of 16 students was able in one month to specify four genetic programs to make groups of *E. coli* cells flash in unison, as fireflies sometimes do. The students did not know how to create DNA sequences, but they had no need to. Endy hired a DNA-synthesis company to manufacture the 58 parts called for in their designs. These new BioBricks were then added to

M.I.T.'s Registry of Standard Biological Parts. That online database today lists more than 140 parts, with the number growing by the month.

Hijacking Cells

AS USEFUL AS IT HAS BEEN to apply the lessons of other fields of engineering to genetics, beyond a certain point the analogy breaks down. Electrical and mechanical machines are generally self-contained. That is true for a select few genetic devices: earlier this year, for example, Milan Stojanovic of Columbia University contrived test tubes of DNA-like biomolecules that play a chemical version of tic-tac-toe. But synthetic biologists are mainly interested in building genetic devices within living cells, so that the systems can move, reproduce and interact with the real world. From a cell's point of view, the synthetic device inside it is a parasite. The cell provides it with energy, raw materials and the biochemical infrastructure that decodes DNA to messenger RNA and then to protein.

The host cell, however, also adds a great deal of complexity. Biologists have invested years of work in computer models of *E. coli* and other single-celled organisms [see "Cybernetic Cells," SCIENTIFIC AMERICAN, August 2001]. And yet, acknowledges Ron Weiss of Princeton, "if you give me the DNA sequence of your genetic system, I can't tell you what the bacteria will do with it." Indeed, Endy recalls, "about half of the 60 parts we designed in 2003 initially couldn't be synthesized because they killed the cells that were copying them. We had to figure out a way to lower the burden that carrying and replicating the engineered DNA imposed on the cells." (Eventually 58 of the 60 parts were produced successfully.)

One way to deal with the complexity added by the cells' native genome is to dodge it: the genetic device can be sequestered on its own loop of DNA, separate from the chromosome of the organism. Physical separation is only half the solution, however, because there are no wires in cells. Life runs on "wetware," with many protein signals simply floating randomly from one part to another. "So if I have one inverter over here made out of proteins and DNA," Endy explains, "a protein signal meant for that part will also act on any other instance of that inverter anywhere else in the cell," whether it lies on the artificial loop or on the natural chromosome.

One way to prevent crossed signals is to avoid using the same part twice. Weiss has taken this approach in constructing a "Goldilocks" genetic circuit, one that lights up when a target chemical is present but only when the concentration is not too high and not too low [see illustration on opposite page]. Tucked inside its various parts are four inverters, each of which responds to a different protein signal. But this strategy makes it much more difficult to design parts that are truly interchangeable and can be rearranged.

Endy is testing a solution that may be better for some systems. "Our inverter uses the same components [as one of Weiss's], just arranged differently," Endy says. "Now the input is not a protein but rather a rate, specifically the rate at which a gene is transcribed. The inverter responds to how many messenger RNAs are produced per second. It makes a protein, and that protein determines the rate of transcription going out [by switching on a second gene]. So I send in TIPS—transcription events per second—and as output, I get TIPS. That is the common currency, like a current in an electrical circuit." In principle, the inverter could be removed and replaced with any other BioBrick that processes TIPS. And TIPS signals are location-specific, so the same part can be used at several places in a circuit without interference.

The TIPS technique will be tested by a new set of genetic systems designed by students who took a winter course at M.I.T. this past January. The aim this year was to reprogram cells to work cooperatively to form patterns, such as polka dots, in a petri dish. To do this the cells must communicate with one another by secreting and sensing chemical nutrients.

"This year's systems were about twice the size of the 2003 projects," Endy says. It took 13 months to get the blinking *E. coli* designs built and into cells. But in the intervening year the inventory of BioBricks has grown, the speed of DNA synthesis has shot up, and the engineers have gained experience assembling genetic circuits. So Endy expects to have the 2004 designs ready for testing in just five months, in time to show off at the first synthetic biology conference, scheduled for this June.

Rewriting the Book of Life

THE SCIENTISTS WHO ATTEND that conference will no doubt commiserate about the inherent difficulty of engineering a relatively puny stretch of DNA to work reliably within a cell that is constantly changing. Living machines reproduce, but as they do they mutate.

"Replication is far from perfect. We've built circuits and seen them mutate in half the cells within five hours," Weiss reports. "The larger the circuit is, the faster it tends to mutate." Weiss and Frances H. Arnold of the California Institute of Technology have evolved circuits with improved performance

BUILDING A GENETIC MACHINE

A living TNT detector that reveals buried land mines could be made using genetic "Goldilocks" circuits that fluoresce only when the concentration of TNT is just right.

CONSTRUCTING A GENETIC TNT DETECTOR

Drawing from interchangeable DNA parts (*in test tubes*), engineers could assemble slightly different circuits. One would glow red, but only when the TNT concentration is high. A second might fluoresce yellow at medium levels of TNT, and a third could glow green at low concentrations.

Engineers would insert the circuits into three separate bacterial cultures. In the soil over a mine (*below*), TNT tapers off in a circular gradient. So a mixture of the altered cells would produce a fluorescent bull's-eye centered on the mine.

Vials of genetic parts

Genetic circuit

Bacterial chromosome

Altered bacterial cells

TNT

Land mine

A TNT DETECTOR IN ACTION

One design for a Goldilocks genetic circuit uses four interacting parts: a sensor, upper and lower thresholds, and an inverter. Each part has a distinctive behavior: the amount of protein output it produces varies as a function of the amount of input protein it receives. In the schematic

below for a red-glowing circuit, the graphs illustrate how each part adjusts its output over the full range of TNT concentrations. (The geometric shapes reflect output levels when the TNT concentration is in the "sweet spot" of, say, 4 percent.)

SENSOR

sends out two signals that are roughly proportional to the level of TNT within the cell. Importantly, the two signals are unequal: at a TNT level of 4 percent, one of the genes in the sensor (*dark purple*) churns out only half as much protein (*squares*) as does the other gene (*light purple*).

UPPER THRESHOLD

receives the weaker signal from the sensor. Output from the first gene in this part starts to fall dramatically but is still high when TNT levels are 4 percent. The next gene in the chain simply inverts whatever signal the first gene generates. So at 4 percent TNT concentration, the upper threshold sends very little protein to the next part (the inverter).

INVERTER

contains genes that express fluorescent proteins only when input signals from both thresholds are low. Its input (*pentagons*) is the sum of the protein signal produced by the lower threshold plus the signal from the latter part of the upper threshold.

The inverter's output—a fluorescent red protein—is high only when its input is low, at TNT concentrations of around 4 percent.

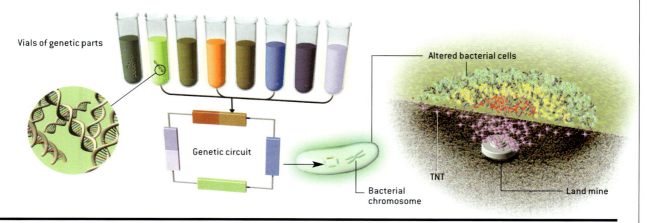

OUTPUT — TNT — 4%

Lower threshold + Upper threshold = Inverter input

OUTPUT — TNT — 4%

TNT

Fluorescent proteins

LOWER THRESHOLD

emits the inverse of its input signal (*triangles*), which is the protein that the sensor produces most prolifically. This part's output begins to fall steeply at TNT levels around 1 percent; by 4 percent TNT, the part produces almost no protein to send to the inverter.

Life, but Not (Exactly) as We Know It

Life on earth has taken a tremendous range of forms, but all species arise from the same molecular ingredients: five nucleotides that form the building blocks for DNA and RNA, and 20 amino acids that serve as building blocks for proteins. (At least two additional natural amino acids are made by a few odd species.) These ingredients limit the chemical reactions that can happen inside cells and so constrain what life can do.

That constraint was eased in 2001, probably for the first time in more than three billion years. After years of trying, Lei Wang, Peter G. Schultz and their co-workers at the Scripps Research Institute in La Jolla, Calif., at last succeeded in adding to *Escherichia coli* bacteria all the genetic components the cells need to decode the three-nucleotide DNA sequence TAG into unnatural amino acids of various kinds.

It was a seminal step but an intermediate one, because amino acids by themselves have relatively few uses. The real goal is to modify cells so that they not only synthesize artificial amino acids but also string them together with natural amino acids to form proteins that no other form of life can make. Last year Schultz's group announced that it had done just that with *E. coli*, and in August the team reported its creation of a similarly talented form of yeast.

"The translational machinery [that reads RNA to make proteins] in yeast is very similar to the translational machinery of

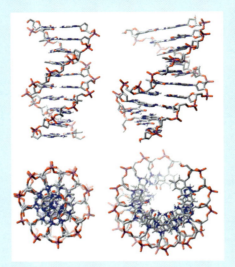

TWISTED LADDER OF DNA (*above left, seen in side view and top view*) may not be the only macromolecule capable of storing the blueprints for living organisms. Scientists are experimenting with semiartificial nucleic acids, such as xDNA (*at right*), that are more stable and thus less likely to suffer mutations.

humans," points out T. Ashton Cropp, a biologist in Schultz's lab. "So far we have produced six kinds of unnatural amino acids in yeast," and the scientists have begun adapting the systems to work in human kidney cells and in roundworms, Cropp says. "We're very close to having a system that can make two different unnatural amino acids and put them in the same protein," he

adds. "It is tricky because in order to do that, the cell has to decode a four-nucleotide DNA sequence," which, as far as anyone knows, no cell has ever done.

"This advance could foster developments with inestimable biomedical potential," suggests Brian L. Davis of the Research Foundation of Southern California in La Jolla. He envisions white blood cells that could make novel proteins to destroy pathogens or cancerous cells more quickly. Cropp says the technology is already producing new research tools, such as proteins that include fluorescent amino acids or that change behavior when they are exposed to light. "It allows us to attach polymers to therapeutic proteins, which makes them work better as drugs," Cropp notes.

Synthetic biologists also have been avidly tinkering with unnatural forms of DNA. Steven A. Benner and his associates at the University of Florida developed a six-letter genetic alphabet more than a decade ago; it was recently used to create a rapid test for the SARS virus. "We're playing around with a variant called TNA, where ribose is replaced with a slightly simpler sugar," says Jack W. Szostak of Massachusetts General Hospital. TNA and xDNA, created by Eric T. Kool of Stanford University, are more stable than DNA. That may make them better suited as a medium for reprogramming cells. First, however, scientists will have to get them working inside living organisms. —*W.W.G.*

---

using multiple rounds of mutation followed by selection of those cells most fit for the desired task. But left unsupervised, evolution will tend to break genetic machines.

"I would like to make a genetically encoded device that accepts an input signal and simply counts: 1, 2, 3, ... up to 256," Endy suggests. "That's not much more complex than what we're building now, and it would allow you to quickly and precisely detect certain types of cells that had lost control of their reproduction and gone cancerous. But how do I design a counter so that the design persists when the machine makes copies of itself that contain mistakes? I don't have a clue. Maybe we have to build in redundancy—or maybe we need to make the function of the counter somehow good for the cell."

Or perhaps the engineers will have to understand better how simple forms of life, such as viruses, have solved the problem of persistence. Synthetic biology may help here, too. Last November, Hamilton O. Smith and J. Craig Venter announced that their group at the Institute for Biological Energy Alternatives had re-created a bacteriophage (a virus that infects bacteria) called phiX174 from scratch, in just two weeks. The syn-

thetic virus, Venter said, has the same 5,386 base pairs of DNA as the natural form and is just as active.

"Synthesis of a large chromosome is now clearly in reach," said Venter, who for several years led a project to identify the minimal set of genes required for survival by the bacterium *Mycoplasma genitalium*. "What we don't know is whether we can insert that chromosome into a cell and transform the cell's operating system to work off the new chromosome. We will have to understand life at its most basic level, and we're a long way from doing that."

Re-creating a virus letter-for-letter does not reveal much about it, but what if the genome were dissected into its constituent genes and then methodically put back together in a way that makes sense to human engineers? That is what Endy and colleagues are doing with the T7 bacteriophage. "We've rebuilt T7—not just resynthesized it but reengineered the genome and synthesized that," Endy reports. The scientists are separating genes that overlap, editing out redundancies, and so on. The group has completed about 11.5 kilobases so far and expects to finish the remaining 30,000 base pairs by the end of 2004.

"The people in this class are happy and building nice, constructive things, as opposed to new species of virus or new kinds of bioweapons."

—Drew Endy, M.I.T.

Beta-Testing Life 2.0

SYNTHETIC BIOLOGISTS have so far built living genetic systems as experiments and demonstrations. But a number of research laboratories are already working on applications. Martin Fussenegger and his colleagues at ETH Zurich have graduated from bacteria to mammals. Last year they infused hamster cells with networks of genes that have a kind of volume control: adding small amounts of various antibiotics turned the output of the synthetic genes to low, medium or high. Controlling gene expression in this way could prove quite handy for gene therapies and the manufacture of pharmaceutical proteins.

Living machines will probably find their first uses for jobs that require sophisticated chemistry, such as detecting toxins or synthesizing drugs. Last year Homme W. Hellinga of Duke University invented a way to redesign natural sensor proteins in *E. coli* so that they would latch onto TNT or any other compound of interest instead of their normal targets. Weiss says that he and Hellinga have discussed combining his Goldilocks circuit with Hellinga's sensor to make land-mine detectors.

Jay Keasling, who recently founded a synthetic biology department at Lawrence Berkeley National Laboratory (LBNL), reports that his group has engineered a large network of wormwood and yeast genes into *E. coli*. The circuit enables the bacterium to fabricate a chemical precursor to artemisinin, a next-generation antimalarial drug that is currently too expensive for the parts of the developing world that need it most.

Keasling says that three years of work have increased yields by a factor of one million. By boosting the yields another 25- to 50-fold, he adds, "we will be able to produce artemisinin-based dual cocktail drugs to the Third World for about one tenth the current price." With relatively simple modifications, the bioengineered bacteria could be altered to produce expensive chemicals used in perfumes, flavorings and the cancer drug Taxol.

Other scientists at LBNL are using *E. coli* to help dispose of nuclear waste as well as biological and chemical weapons. One team is modifying the bacteria's sense of "smell" so that the bugs will swim toward a nerve agent, such as VX, and digest it. "We have engineered *E. coli* and *Pseudomonas aeruginosa* to precipitate heavy metals, uranium and plutonium on their cell wall," Keasling says. "Once the cells have accumulated the metals, they settle out of solution, leaving cleaned wastewater."

Worthy goals, all. But if you become a touch uneasy at the thought of undergraduates creating new kinds of germs, of private labs synthesizing viruses, and of scientists publishing papers on how to use bacteria to collect plutonium, you are not alone.

In 1975 leading biologists called for a moratorium on the use of recombinant-DNA technology and held a conference at the Asilomar Conference Grounds in California to discuss how to regulate its use. Self-policing seemed to work: there has yet to be a major accident with genetically engineered organisms. "But recently three things have changed the landscape," Endy points out. "First, anyone can now download the DNA sequence for anthrax toxin genes or for any number of bad things. Second, anyone can order synthetic DNA from offshore companies. And third, we are now more worried about intentional misapplication."

So how does society counter the risks of a new technology without also denying itself all the benefits? "The Internet stays up because there are more people who want to keep it running than there are people who want to bring it down," Endy suggests. He pulls out a photograph of the class he taught last year. "Look. The people in this class are happy and building nice, constructive things, as opposed to new species of virus or new kinds of bioweapons. Ultimately we deal with the risks of biological technology by creating a society that can use the technology constructively."

But he also believes that a meeting to address potential problems makes sense. "I think," he says, "it would be entirely appropriate to convene a meeting like Asilomar to discuss the current state and future of biological technology." This June, as leaders in the field meet to share their latest ideas about what can now be created, perhaps they will also devote some thought to what shouldn't. SA

W. Wayt Gibbs is senior writer.

MORE TO EXPLORE

An Expanded Eukaryotic Genetic Code. Jason W. Chin et al. in *Science*, Vol. 301, pages 964–967; August 15, 2003.

Genetic Circuit Building Blocks for Cellular Computation, Communications, and Signal Processing. Ron Weiss et al. in *Natural Computing*, Vol. 2, No. 1, pages 47–84; 2003.

The M.I.T. Synthetic Biology Working Group: **syntheticbiology.org**

The M.I.T. Registry of Standard Biological Parts: **parts.mit.edu**

Article Review: CELL BIOLOGY

Synthetic Life   by W. Wayt Gibbs

TEST YOUR COMPREHENSION

1. The terminology of synthetic biology is a mixture of genetic terms and engineering terms. Develop a list of genetic terms related to the regulation of genetic circuits and match the genetic terms with engineering terms related to circuits.

2. The BioBricks described in this article are carried on circular plasmids that can be readily engineered with interchangeable components. What are the key characteristics of gene structure and gene expression that might be important to regulate in this type of system?

   ANSWER: *The regulation of gene expression can be at the level of transcriptional regulation (mediated by the promoter in front of the gene itself) or at the level of translation. Using interchangeable promoter elements, which can be induced by various environmental conditions or media additives, is quite effective for controlling the expression of a gene in an engineered plasmid.*

3. The author mentions that ~50% of the initial elements developed in 2003 turned out to kill the host E. coli cells, and thus were not useful until revised versions were developed. What is the author's explanation for the lethality of many of the initially developed BioBrick components?

   ANSWER: *The regulatory elements being developed likely interfered with the essential biological functions of the living bacterium, and thus could not be tolerated within that host cell.*

CLASS ACTIVITIES AND DISCUSSION

1. This article mentions "artificial" amino acids being used to make proteins. This refers to making proteins from amino acids other than the 20 used in naturally existing cells. What is the basic backbone of any amino acid? What component of the amino acid could be varied to make unique protein structures?

2. Examine the "blinking" E. coli described in this manuscript, exhibiting periodic bursts of expression of the green fluorescent protein (GFP). What is the fundamental timing component of this circuit? What inherent cellular phenomenon is continually being activated and then inactivated by this regulatory circuit?

3. The BioBricks registry now has hundreds of components available for building biological circuits. Divide the class into 6 groups. Each group will develop a proposal for an interesting "genetically engineered machine" or GEM. What types of applications might you predict for your GEM, were it to be built?

4. The TNT sensing bacterium proposed in this article depends upon a Goldilocks genetic circuit built upon an initial "sensor". If you were a genetic engineer working on this project, where would you look for a protein or genetic element that responds quantitatively to the concentration of TNT? If you identified a protein that responded to a molecule similar to TNT, how might you begin to isolate mutant forms of this protein that bound and responded to TNT instead?

Article Review: GENETICS

TESTING YOUR COMPREHENSION

1. Drew Endy and his colleagues at MIT are reengineering the genome of the T7 bacteriophage, synthesizing a new genome in the laboratory. What types of changes are they placing into that genome?

   ANSWER: *The researchers mention separating overlapping genes and removing redundant genes. This reengineering of a viral into a genome that makes sense to our understanding of genome function will certainly reveal some surprises about genome function.*

2. What was the purpose of the conference held at the Asilomar Conference Grounds in 1975?

   ANSWER: *In the early days of recombinant DNA studies, scientists became concerned that some research could place the public at risk. The conference attendees developed guidelines for ethical and safe experimentation in this new area of work. The conference is still heralded as an important success, and some feel it is now time for a similar synthetic biology conference.*

3. The work of Steven Benner includes the development of a six-letter genetic alphabet, while Jack Szostak and Eric Tool have developed modified forms of DNA. Describe the challenges to using these new forms of DNA in a living cell.

   ANSWER: *The DNA of a living cell carries a structure that is efficiently transcribed by cellular RNA polymerase, and the RNA generated is read and translated by the ribosome. For an alternate genetic code or form of DNA to function within a living cell, many cellular components would have to be adapted to facilitate this new code — a significant challenge.*

CLASS ACTIVITIES AND DISCUSSION

1. In response to the rapid development and progression of synthetic biology as a new research area in genetic engineering, some scientists have raised a voice of caution. The editors of Nature offered comment in 2004, stating, " Researchers must do more than talk among themselves. They must demonstrate publicly that they are willing to consult and reflect carefully about risk – both perceived and genuine – and to moderate their actions accordingly." Do you agree with this commentary? What types of abuse might progress in synthetic biology facilitate?

   Futures of Artificial Life (2004), *Nature* 431:613.

2. An early version of the circuits described here is known as the "repressilator" circuit developed by Michael Elowitz and Stan Liebler (Princeton University) to generate an oscillating output signal. Define the nature of this circuit and the genetic source of the three essential components of the circuit.

   Elowitz MB, and S. Leibler (2000), A synthetic oscillatory network of transcriptional regulators, *Nature* 403: 335-8.

3. In 2004, undergraduate students from 5 universities gathered for the Synthetic Biology Jamboree (now renamed as the annual intercollegiate Genetically Engineered Machine competition or iGEM). A team of undergraduates from the University of Texas at Austin constructed light responsive bacteria, generated in collaboration with Christopher Voight's research group (University of California, San Francisco). Thus a plate of bacteria act as a crude photographic film changing their color (black/white) in response to light exposure. Examine this reference manuscript and define the circuit they used. Brainstorm on possible projects your class could develop for a similar competition.

   Leveskya (2005), Engineering *Escherichia coli* to see light, *Nature* 438: 441-2.

   Campbell AM (2005), Meeting Report: Synthetic Biology Jamboree for Undergraduates, *Cell Biology Education* 4:19-23.

   iGEM web link:
   http://parts2.mit.edu/wiki/index.php/Main_Page

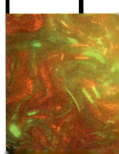

Hitting the Genetic OFF Switch

A host of start-ups is speeding development of a new class of drugs that block the action of RNA   By Gary Stix

In 1996 *Worth* magazine proclaimed that Isis Pharmaceuticals could become the next Microsoft, a prediction that turned out to be a particularly egregious example of hyperbole run amok. To be sure, Isis remains a leader in the gene-blocking technology called antisense. But the road to successful treatments for cancer and other diseases has been littered with disappointments.

During the past few years, a new gene-silencing technology has emerged that may be poised to fulfill the promise that was once trumpeted for antisense. "I've been writing in grants for 25 years that during the next five years I'm going to test this process or that process to see if I can do gene inactivation studies in mammalian cells in culture. And I did them, and they were so awkward and so complicated that you just couldn't apply them generally," says Phillip A. Sharp, director of the McGovern Institute for Brain Research at the Massachusetts Institute of Technology. "Lo and behold, all of the time right there in front of me was a process that I could have used."

Sharp, a co-winner of the 1993 Nobel Prize in Physiology or Medicine, was referring to a series of relatively recent discoveries that cells have a mechanism, dubbed RNA interference (RNAi), which blocks gene expression. It prevents RNA transcripts of genes from giving rise to the proteins those genes encode. This natural method of gene silencing comes into play, for example, when viruses try to commandeer a cell's protein-making machinery to produce viral proteins.

A milestone arrived in 1998, when Andrew Z. Fire, now at the Stanford University School of Medicine, and Craig C. Mello of the University of Massachusetts Medical School identified in worms double-stranded RNAs that acted as the switch to turn off genes in RNAi. And in 2001 Thomas Tuschl, now at the Rockefeller University, found that an abbreviated version of double-stranded RNAs—short interfering RNAs (siRNAs)—could shut off genes in mammalian cells. The number of research papers on RNAi has mushroomed from a dozen-plus in 1998 to multiple hundreds last year. Even if the promise for therapeutics never materializes, it is quite likely that some of the seminal discoveries will garner Nobel Prizes. "This has touched everything we do in biological science, from plants to man," Sharp notes. [See "Censors of the Genome," by Nelson C. Lau and David P. Bartel; Scientific American, August 2003.]

The excitement about siRNAs as drugs relates to how they differ in critical ways from antisense therapeutics. At first glance, siRNAs seem very similar to antisense. An antisense drug consists of an artificially synthesized chain of nucleotides, or genetic building blocks, that binds to a messenger RNA containing a complementary sequence. This binding blocks gene expression. An siRNA also silences genes—and it even uses a complementary RNA, or antisense, strand to do so. Once inside a cell, an siRNA attaches to an aggregate of proteins called an RNA-induced silencing complex (RISC), which retains only the antisense strand. The siRNA-bearing RISC then binds to the targeted messenger RNA and degrades it or prevents it from functioning [*see box on page 58*].

Unlike the antisense drugs that have been under development for the past 15 years, siRNAs do not disrupt only a single messenger RNA. They act as catalysts, doing the same job over and over, one explanation for their apparent potency. "They are 100- to 1,000-fold more effective than antisense," says Judy Lieberman, a senior investigator at the CBR Institute for Biomedical Research in Boston and one of the first researchers to show the therapeutic potential of the technique in animals.

Already almost 100 companies are

involved in RNAi; nearly half supply the chemicals and technology needed to perform experiments, and the others are biotechnology or pharmaceutical companies doing commercial research with RNAi, according to Kewal K. Jain, chief executive officer of Jain PharmaBiotech, a Basel, Switzerland, market research firm. "All of this has happened within the last two or three years," Jain says.

A small fraction of these companies have dedicated themselves to producing therapies using siRNAs. As soon as Tuschl's paper documenting siRNAs in mammalian cells was published, the venture-capital community sprang into action. "It was worth it to make a bet realizing in vivo efficacy was not guaranteed," says Christoph H. Westphal, one of the

Paul R. Schimmel, a professor of molecular biology and chemistry at the Scripps Research Institute, and the founder of several biotechnology companies before this one, insisted on the name Alnylam, an Arabic word meaning "string of pearls" that is also the designation for the middle star of Orion's belt. Schimmel made the case, over the protests of others, that the name—pronounced "al-NIGH-lam"—was difficult to pronounce but impossible to forget. Barry Greene, the company's chief operating officer, furnishes a simpler explanation: "The URL was open," he joked at a recent investors' conference.

The founders constituted an all-star scientific advisory team, and some also filled slots on the board of directors. But

scientific smarts, would determine who would thrive or falter as drug development and clinical trials got under way. "We were very focused and running very hard," he remembers. "We recognized that if we weren't first, others would grab it from us."

The fledgling Alnylam even bought the German firm Ribopharma to get a hold of a key patent. The stir created by RNAi—tagged by *Science* magazine as "breakthrough of the year" in 2002—helped to bring in venture money. The total take thus far has reached about $85 million, including a somewhat disappointing initial public offering in the spring, and provides enough to keep Alnylam going for another two years, until the first drug makes it through the preliminary safety phase of clinical trials.

The success or failure of RNAi as a therapeutic will hinge on getting the drug into target cells without its being chopped up by enzymes. The drug must then persist in the cell long enough to carry out its job of binding to and inhibiting specific messenger RNAs. The challenge of delivery and stabilization has also posed a significant hurdle for the success of antisense.

These difficulties serve as one reason why the newly formed Alnylam team immediately discarded one approach to delivery: using a vector—a virus, for instance—to ferry a stretch of DNA, not just past the cell wall, but into the nucleus. The gene would then go on to make the RNA that would interfere with gene expression. "In my mind, nothing about RNAi solves the problem of gene therapy," Maraganore notes, referring to the downsides of using viruses to deliver the drug to the right location and the unwanted side effects that they sometimes provoke. Consequently, short interfering RNAs are synthesized in the laboratory from a soup of nucleotides until they form double-stranded molecules that have 21 nucleotide pairs. Some other companies, such as Benitec in Australia, are still pursuing a gene therapy approach [see table on page 59].

Key expertise and intellectual property to accomplish this task came from an unlikely source. Alnylam had hired away

BINDING A MESSENGER RNA (*long strand*) to complementary RNA and an aggregate of proteins (*blob*) could potentially become a lucrative new approach to drug development.

founders of Alnylam in Cambridge, Mass., and a general partner with Polaris Venture Partners. Many of the early innovators in RNAi technology, including Tuschl, Sharp and David P. Bartel of M.I.T., among others, got together to form Alnylam Pharmaceuticals in 2002. Sharp, a founder of biotech giant Biogen, brought together this banner group after conversations with more established companies failed to generate sufficient interest.

when John M. Maraganore, the company's first permanent chief executive, a transplant from Millennium Pharmaceuticals, hired the people who actually would be entrusted with the task of making siRNA drugs, he did not at first seek out postdoctoral students of these research heavyweights. "Five people were focused on intellectual property and one on science," Maraganore says. A bulletproof patent estate, as much as

from Isis an executive, Muthiah Manoharan, to become vice president of drug discovery. Maraganore called Isis president Stanley T. Crooke last summer and reassured him that Alnylam still wished to be on good terms with the antisense manufacturer. A few months later a dialogue between the two companies resulted in an agreement in which Alnylam would pay $5 million to license Isis's ex-tensive patent portfolio of chemical techniques for delivering and stabilizing RNA. "We will be able to take advantage of the 10-plus years of development of chemistry used in antisense," Maraganore says. In turn, Isis invested $10 million in Alnylam, giving it a 5 percent equity stake in the company and a stream of royalties and fees once siRNA products hit the marketplace. It will also get the rights to make some siRNA drugs.

The development trajectory for siRNA recapitulates the path that antisense has taken. The only antisense drug approved to date is Isis's Vitravene, intended to treat an eye disease once prevalent in AIDS patients. The drug is injected directly into the eye, concentrating the compound at its target while impeding it from producing adverse side effects in

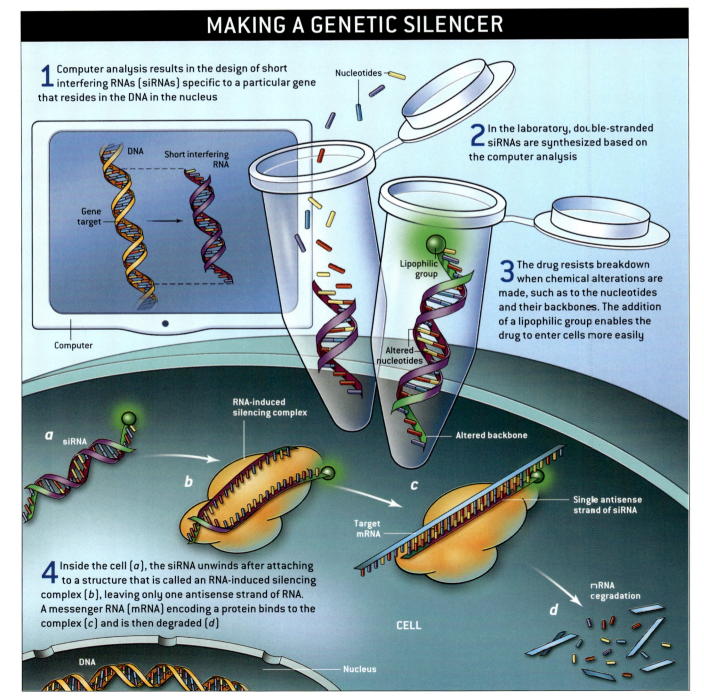

MAKING A GENETIC SILENCER

1 Computer analysis results in the design of short interfering RNAs (siRNAs) specific to a particular gene that resides in the DNA in the nucleus

Nucleotides

2 In the laboratory, double-stranded siRNAs are synthesized based on the computer analysis

DNA
Short interfering RNA
Gene target
Computer

Lipophilic group

3 The drug resists breakdown when chemical alterations are made, such as to the nucleotides and their backbones. The addition of a lipophilic group enables the drug to enter cells more easily

Altered nucleotides

RNA-induced silencing complex

a siRNA

b

c

Altered backbone

Single antisense strand of siRNA

Target mRNA

4 Inside the cell (*a*), the siRNA unwinds after attaching to a structure that is called an RNA-induced silencing complex (*b*), leaving only one antisense strand of RNA. A messenger RNA (mRNA) encoding a protein binds to the complex (*c*) and is then degraded (*d*)

mRNA degradation

d

CELL

DNA

Nucleus

BIOTECHS DEVELOPING DRUGS BASED ON RNAi

COMPANY	THERAPIES
Acuity Pharmaceuticals Philadelphia	Age-related macular degeneration and diabetic retinopathy
Alnylam Pharmaceuticals Cambridge, Mass.	Age-related macular degeneration, Parkinson's disease and, over the longer term, cancer, metabolic and autoimmune diseases
atugen Berlin	Cancers and metabolic diseases, for systemic applications, and ocular and skin diseases, for topical delivery
Benitec Queensland, Australia	Hepatitis C and, over the longer term, cancer, autoimmune and viral diseases using a gene therapy approach
CytRx Los Angeles	Amyotrophic lateral sclerosis, cytomegalovirus, obesity and type 2 diabetes
Intradigm Rockville, Md.	Cancer
Nucleonics Horsham, Pa.	Hepatitis B and C and, over the long run, cancer, inflammatory and viral diseases using a gene therapy approach
Sirna Therapeutics Boulder, Colo.	Macular degeneration, hepatitis C and, later, cancer, metabolic, inflammatory, dermatological and central nervous system diseases

other parts of the body. But the market for the drug virtually disappeared as other treatments for AIDS became available and prevented most cases of the cytomegalovirus retinitis infection.

Short RNAs may eventually be delivered to the bloodstream to treat systemic diseases. But, reprising the antisense experience, the first petition to begin a clinical trial was filed in August by Acuity Pharmaceuticals, a Philadelphia company that will attempt to treat age-related macular degeneration by intraocular injection. Other companies, including Alnylam, will follow suit with their own drug trials for macular degeneration. One of these filings will bring Alnylam head to head with the other drug developer that stands a chance of becoming a leader in this emerging market niche.

A Boulder, Colo., company called Sirna Therapeutics was expected to submit an application to the FDA for a macular degeneration drug in early September, perhaps half a year or more before Alnylam. Unlike Alnylam, Sirna is no start-up. It is a reincarnation of another firm, Ribozyme Pharmaceuticals, which for a decade staked its fate on a different type of RNA-related drug. Ribozymes are RNAs acting as catalytic enzymes that, in principle, can cut up messenger RNA and prevent a protein from being produced. But, as with antisense, the potency of ribozymes came into question. A drug to combat hepatitis C caused a monkey to go blind, probably because of the massive doses injected. And another drug did not seem to slow growth of tumors in patients with advanced breast cancer. At the time, Howard W. Robin, a new chief executive, who had managed development of drugs like Betaseron for multiple sclerosis at Berlex Laboratories, was faced with a decision about whether to close shop. The company had only $2 million in cash left and risked being delisted from the Nasdaq.

The darkest days coincided with Tuschl's publication about RNAi in mammalian cells. Instead of turning out the lights, Ribozyme Pharmaceuticals became Sirna Therapeutics. Sirna rejiggered the chemical techniques it had used to deliver and stabilize ribozymes and adapted them to siRNAs. Robin claims that single doses of its siRNAs have remained active inside cells of live animals for up to 22 days. The revamping succeeded in attracting $72 million in new investment during an 18-month period. Besides macular degeneration, the company has programs in hepatitis, oncology and Huntington's disease, among others. "It's not often that you take the skills and intellectual property from a technology that's not working very well and transfer it to the hottest area of biology," Robin says.

Sirna has filed for 90 patents that Robin believes cover the most attractive drug prospects. Patent fights may loom as the technology gets nearer the marketplace. "If you look at our competitors, we believe almost everything they're doing violates our patents," Robin proclaims. But Maraganore begs to differ: "We have a bit of a toll road for anyone doing therapeutics." Any dispute is not likely to emerge until siRNA drugs are much closer to approval. In the meantime, investigators will be closely watching whether siRNAs produce unwanted immune responses or shut down genes they are supposed to leave intact.

For the time being, optimism about RNAi reigns. "If you do with RNAi in man what you do in cell culture, you have the most unbelievably powerful technology for making pharmaceuticals," Maraganore says. "It's the dream of medicine to do selective and efficient gene silencing," Robin asserts. But Isis's Crooke, tempered by the failure of trials for a few of his company's antisense drugs, has a slightly different perspective: "Any time you think something is magic, you're going to get in trouble. [RNAi] is a complicated system with lots of interesting nuances that mechanistically should lead to some unexpected effects as well as those you desire."

RNAi has become a preeminent research tool in a remarkably short time. But its potential as a genetically based pharmaceutical will not become clear for several years, when the first clinical trials prove whether a simple injection is capable of shutting down the effects of a disease-causing gene. **SA**

MORE TO EXPLORE

The RNAi Revolution. Carl D. Novina and Phillip A. Sharp in *Nature,* Vol. 430, pages 161–164; July 8, 2004.

The Silent Revolution: RNA Interference as Basic Biology, Research Tool and Therapeutic. Derek M. Dykxhoorn and Judy Lieberman in *Annual Review of Medicine,* Vol. 56 (in press).

Hitting the Genetic OFF Switch *by Gary Stix*

TEST YOUR COMPREHENSION

1. RNA interference is a natural cellular defense mechanism to viral infection. Explain how this might work as a defense mechanism.

   ANSWER: *A virus takes over a newly infected cell by using cellular transcriptional and translational machinery to express its own genes. Once the viral proteins have been synthesized, the cell is converted into a virus manufacturing facility. Cells can defend against this process by identifying foreign RNAs, and destroying them before they are translated.*

2. What is the clear technical hurdle that pharmaceutical companies must overcome if siRNA is to become an effective therapy for human disease?

   ANSWER: *The delivery of siRNA to the desired cells is the central difficulty with this technology. In cultured cells, it is easy to deliver the siRNA into the cells, but with an intact organism, it is a much more difficult task.*

3. Define the prominent mechanisms for delivery of siRNAs to living cells. How do researchers deliver the siRNA get into a living cell?

   ANSWER: *There are a diversity of strategies, including packaging the siRNAs into lipid enclosed micelles, attaching a lipophilic compound directly to the siRNA, generating a viral DNA that expresses the siRNA and infecting cells with that virus, or even injecting the material directly into a target organ (intraocular injection).*

4. In plants, RNAi activity can move from one plant cell to another via the plasmodesmata. How might this benefit the entire plant?

   ANSWER: *Plants can gain an RNAi-mediated resistant to specific plant viruses as infected cells share viral RNA fragments with other cells, generating a RISC enzyme-mediated immunity throughout the plant.*

CLASS ACTIVITIES AND DISCUSSION

1. One critical detail for each of the companies developing RNAi therapies is the specific gene target whose protein product causes the disease being treated, and whose destruction would "cure" the patient without harming the patient What types of genes might be abnormally activated to initiate or sustain a specific disease state (telomerase and cancer, etc.)? Develop a list. Might this technology be used to fight chronic viral infections? Explain.

2. This article highlights some key collaborative efforts among biotech and pharmaceutical companies. Collaboration has long been a key tenet of successful research in academia, but is not often observed among competing companies. Define the key examples of collaboration and mutual benefit offered by this article. Are their hints at problems to come once an effective therapy is developed? Discuss the comments made about patents as they relate to the development of RNAi as a clinical therapy. Discuss the pros/cons of competition and collaboration as it relates to the development of new therapeutic technology.

3. There are a variety of companies that now offer reagents and support for RNAi-based laboratory studies. Examine the website of a company named: Ambion: The RNA Company. Use this website to compare/contrast the RNAi methodology used for mammalian cells and the methodology used for non-mammalian cells (*Caenorhabditis elegans* – roundworm, *Drosophila melanogaster* – fruit fly). Examine the resources available from this company and related companies.

Article Review: GENETICS

TESTING YOUR COMPREHENSION

1. Define the RISC enzyme and the associated enzymatic activity. How is this enzyme complex capable of destroying specific RNA target sequences?

   ANSWER: *The RISC enzyme (RNA-induced silencing complex) is made up of protein and antisense RNA from the target RNA (ds viral RNA, siRNA, etc.). The RNA component of the RISC enzyme directs its RNA-cutting activity at RNA fragments that are complementary to its sequence. The RISC enzyme is stable and can destroy thousands of RNAs, making it extremely effective.*

2. Differentiate siRNA technology from the initial "anti-sense RNA" strategy for interrupting the expression of specific genes.

   ANSWER: *siRNA strategies confer the stable destruction of mRNAs that are recognized by the RISC enzyme. Antisense strategies involve the addition of a synthetic strand of nucleic acid that complements a specific mRNA, blocking its translation.*

3. Many transposons (mobile genetic elements) produce a double-stranded RNA structure as a part of their movement from one genetic locus to another. Might RNAi mechanisms be a defense mechanism against transposons?

   ANSWER: *Yes, geneticists have theorized that RNAi may serve to battle transposons as well as viruses. Transposons cause mutation during chromosomal integration.*

4. Ribozyme Pharmaceuticals was founded to develop therapeutically active RNA that could be delivered to the cells of patients. The apparent lack of potency of these RNAs threatened the viability of the original company. The company was reorganized, renamed Sirna Therapeutics, and is currently developing RNAi-based strategies for their target genes. What key component of their ribozyme work has been adapted to benefit their RNAi-based therapeutics?

   ANSWER: *This company adapted their strategies for delivery and stabilization of ribozymes to their new work with RNA.*

CLASS ACTIVITIES AND DISCUSSION

1. RNAi offers a powerful research tool to reduce (or eliminate) the production of a specific protein in living cells. Examining the phenotype (trait) of cells that lack that specific protein offers valuable insight into the role of that protein in normal cellular function. Before RNAi methodology, what other techniques were available for the generation of cells that lack a specific protein? Discuss the technical challenges of each strategy relative to RNAi.

2. Beyond the development of RNAi as a therapeutic strategy, there have been a several discoveries linking the small RNAs involved in RNAi to genetic phenomenon. One collaborative group revealed that RNAi plays a central role in cellular epigenetics (patterns of gene expression that are inherited independent of gene sequence). Researchers have determined that small RNAs play a key role in the regulation of heterochromatin/euchromatin as cells divide (Robert Martienssen – Cold Spring Harbor Laboratory, Shiv Grewal – National Cancer Institute). Define heterochromatin and euchromatin and relate these forms of chromosome structure to gene expression.

3. One component of RNAi that is not discussed in this article is an enzyme known as "dicer". Research the function of this enzyme as it relates to the RISC enzyme complex. Define the activity of dicer, its specificity for substrate, and the role it plays in activating the RISC activity. As a component of your research, examine micro RNAs (miRNAs). How do these relate to the RNAi mechanism?

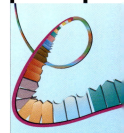

The old axiom "one gene, one protein" no longer holds true. The more complex an organism, the more likely it became that way by extracting multiple protein meanings from individual genes

The Alternative Genome

BY GIL AST

GENES of mice and men are 88 percent alike. Many of the ways that humans differ from rodents arise from how we edit our genetic information.

Spring of 2000 found molecular biologists placing dollar bets,

trying to predict the number of genes that would be found in the human genome when the sequence of its DNA nucleotides was completed. Estimates at the time ranged as high as 153,000. After all, many said, humans make some 90,000 different types of protein, so we should have at least as many genes to encode them. And given our complexity, we ought to have a bigger genetic assortment than the 1,000-cell roundworm, *Caenorhabditis elegans,* which has a 19,500-gene complement, or corn, with its 40,000 genes.

When a first draft of the human sequence was published the following summer, some observers were therefore shocked by the sequencing team's calculation of 30,000 to 35,000 protein-coding genes. The low number seemed almost embarrassing. In the years since, the human genome map has been finished and the gene estimate has been revised downward still further, to fewer than 25,000. During the same period, however, geneticists have come to understand that our low count might actually be viewed as a mark of our sophistication because humans make such incredibly versatile use of so few genes.

Through a mechanism called alternative splicing, the information stored in the genes of complex organisms can be edited in a variety of ways, making it possible for a single gene to specify two or more distinct proteins. As scientists compare the human genome to those of other organisms, they are realizing the extent to which alternative splicing accounts for much of the diversity among organisms with relatively similar gene sets. In addition, within a single organism, alternative splicing allows different tissue types to perform diverse functions working from the same small gene assortment.

Indeed, the prevalence of alternative splicing appears to increase with an organism's complexity—as many as three quarters of all human genes are subject to alternative editing. The mechanism itself probably contributed to the evolution of that complexity and could drive our further evolution. In the shorter term, scientists are also beginning to understand how faulty gene splicing leads to several cancers and congenital diseases, as well as how the splicing mechanism can be used therapeutically.

Pivotal Choices

THE IMPORTANCE of alternative editing to the functioning of many organisms cannot be overestimated. For example, life and death depend on it—at least when a damaged cell must determine whether to go on living. Each cell constantly senses the conditions inside and outside itself, so that it can decide whether to maintain growth or to self-destruct in a preprogrammed process known as apoptosis. Cells that cannot repair DNA will activate their apoptotic program. Craig B. Thompson of the University of Pennsylvania and his colleagues have recently shown that a gene called *Bcl-x,* which is a regulator of apoptosis, is alternatively spliced to produce either of two distinct proteins, Bcl-x(L) and Bcl-x(S). The former suppresses apoptosis, whereas the latter promotes it.

The initial discovery that cells can give rise to such different forms of protein from a single gene was made some 25 years ago, but the phenomenon was considered rare. Recent genome comparisons have revealed it to be both common and crucial, adding a dramatic new twist to the classical view of how information

Overview/*Cut-and-Paste Complexity*

- A gene's instructions can be edited by cellular machinery to convey multiple meanings, allowing a small pool of protein-coding genes to give rise to a much larger variety of proteins.
- That such alternative splicing of genetic messages is possible was long understood. But only when the genome sequences of humans and other organisms became available for side-by-side comparison did geneticists see how widespread alternative splicing is in complex organisms and how much the mechanism contributes to differentiating creatures with similar gene sets.
- Alternative splicing enables a minimal number of genes to produce and maintain highly complex organisms by orchestrating when, where and what types of proteins they manufacture. Humans, in turn, may soon be able to regulate our own gene splicing to combat disease.

ONE GENE, MANY PROTEINS

The classical view of gene expression was simple: a DNA gene is first transcribed into RNA form, then cellular splicing machinery edits out "junk" stretches called introns and joins meaningful portions called exons into a final messenger RNA (mRNA) version, which is then translated into a protein. As it turns out, these rules do not always apply. In complex organisms, the initial RNA transcript can be alternatively spliced—exons may be discarded and introns, or portions of them, retained—to produce a variety of mRNAs, and thus different proteins, from a single gene.

CLASSIC GENE EXPRESSION

A DNA sequence is transcribed into a single-stranded copy made of RNA. Cellular machinery then "splices" this primary transcript: introns—each of which is defined by distinctive nucleotide sequences at its beginning and end, known, respectively, as the 5′ (five-prime) and 3′ (three-prime) splice sites—are removed and discarded while exons are joined into an mRNA version of the gene that will be translated into a protein by the cell.

ALTERNATIVE SPLICING

A gene's primary transcript can be edited in several different ways, shown at the right, where splicing activity is indicated by dashed lines. An exon may be left out (*a*). Splicing machinery may recognize alternative 5′ splice sites for an intron (*b*) or alternative 3′ splice sites (*c*). An intron may be retained in the final mRNA transcript (*d*). And exons may be retained on a mutually exclusive basis (*e*).

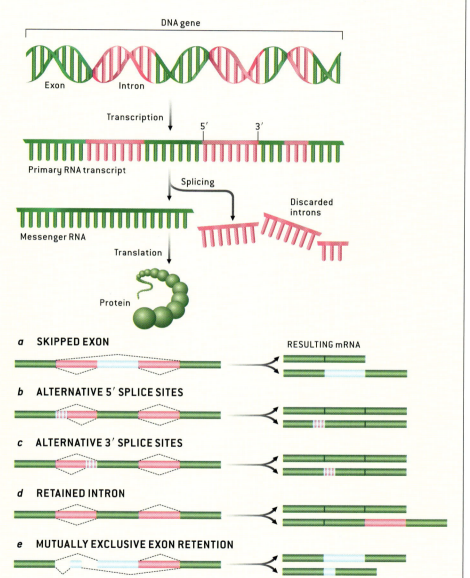

DNA gene

Exon   Intron

Transcription

Primary RNA transcript

Splicing

Messenger RNA   Discarded introns

Translation

Protein

a   **SKIPPED EXON**   RESULTING mRNA

b   **ALTERNATIVE 5′ SPLICE SITES**

c   **ALTERNATIVE 3′ SPLICE SITES**

d   **RETAINED INTRON**

e   **MUTUALLY EXCLUSIVE EXON RETENTION**

■ Exon always spliced in
□ Exon alternatively spliced
■ Intron

stored in a gene is translated into a protein. Most of the familiar facts still hold true: whole genomes contain all the instructions necessary for making and maintaining an organism, encoded in a four-letter language of DNA nucleotides (abbreviated A, G, C and T). In human chromosomes, roughly three billion nucleotides are strung together on each of two complementary strands that form a double helix. When the time comes for a gene's instructions to be "expressed," the double-stranded zipper of DNA opens just long enough for a single-stranded copy of the gene's sequence to be manufactured from a chemical cousin, RNA. Each sequence of DNA nucleotides that gets transcribed into an RNA version in this manner is called a gene. Some of the resulting RNA molecules are never translated into proteins but rather go on to perform housekeeping and regulatory functions within the cell [see "The Unseen Genome: Gems among the Junk,"

THE SPLICING MACHINE

Once a primary RNA transcript of a gene has been created, a structure called the spliceosome carries out RNA editing. In complex organisms, this process is controlled by splicing regulatory (SR) proteins that define exons and direct the spliceosome to specific splice sites. These regulatory molecules therefore determine when and how alternative splices of a gene will be generated. SR proteins are themselves produced in varying forms in different tissues and cell types or during different stages of development within the same tissue.

EXON DEFINITION

An SR protein binds to each exon in the transcript at a distinctive nucleotide sequence called an exonic splicing enhancer (ESE). The SR protein's binding defines the exon for the splicing machinery by recruiting small nuclear RNA (snRNA) molecules called U1 and U2 to splice sites on adjacent introns.

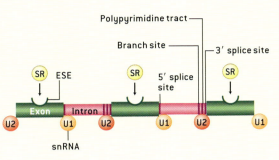

SPLICEOSOME FORMATION

When the original snRNAs have recognized the intron's splice sites, they form a complex with additional snRNAs and more than 100 proteins. This spliceosome complex snips out the introns and joins the exons to produce the mature mRNA.

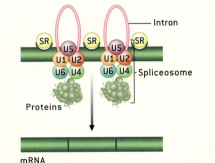

SPLICING SUPPRESSION

An SR protein may also suppress rather than enhance the binding of snRNAs, in which case the sequence to which it binds is called an exonic splicing suppressor (ESS). The SR protein can thus cause an exon to be spliced out of the final mRNA. In humans and other mammals, such exon skipping is the most prevalent form of alternative splicing.

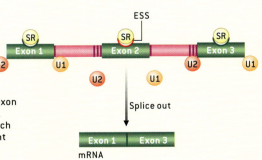

by W. Wayt Gibbs; SCIENTIFIC AMERICAN, November 2003]. The RNA transcripts of genes that do encode a protein will ultimately be read by cellular machinery and translated into a corresponding sequence of amino acids. But first the preliminary transcript must undergo an editing process.

In 1977 Phillip A. Sharp of the Massachusetts Institute of Technology and Richard J. Roberts of New England Biolabs discovered that these initial, or primary, RNA transcripts are like books containing many nonmeaningful chapters inserted at intervals within the text. The nonsense chapters, called introns, must be excised and the meaningful chapters connected together for the RNA to tell a coherent story. In the cutting-and-ligation process, known as splicing, the introns are snipped out of the primary transcript and discarded. Segments of the transcript containing meaningful protein-coding sequences, called exons, are joined together to form a final version of the transcript, known as messenger RNA (mRNA) [see box on preceding page].

But by 1980 Randolph Wall of the University of California at Los Angeles had shown that this basic view of pre-mRNA splicing, in which all introns are always discarded and all exons are always included in the mRNA, does not invariably hold true. In fact, the cellular machinery can "decide" to splice out an exon or to leave an intron, or pieces of it, in the final mRNA transcript. This ability to alternatively edit pre-mRNA transcripts can significantly increase any gene's versatility and gives the splicing mechanism tremendous power to determine how much of one type of protein a cell will produce over the other possible types encoded by the same gene.

In 1984 Tom Maniatis, Michael Green and their colleagues at Harvard University developed a test-tube procedure to reveal the molecular machinery that performs the cutting of introns and pasting together of exons. Details of its workings, and of the regulatory system controlling it, are still being filled in, but this research is unveiling an exquisitely intricate system with fascinating origins.

The Splicing Machine

IN COMPLEX ORGANISMS, two distinct levels of molecular equipment are involved in splicing pre-mRNA transcripts. The so-called basal machinery, which is found in all organisms whose genomes contain introns, has been highly conserved through evolutionary time, from yeast to humans. It consists of five small nuclear RNA (snRNA) molecules, identified as U1, U2, U4, U5 and U6. These molecules come together with as many as 150 proteins to form a complex called the spliceosome that is responsible for recognizing the sites where introns begin and end, cutting the introns out of the pre-mRNA transcript and joining the exons to form the mRNA.

Four short nucleotide sequences

within introns serve as signals that indicate to the spliceosome where to cut [*see box on opposite page*]. One of these splicing signals sits at the beginning of the intron and is called the 5' (five-prime) splice site; the others, located at the end of the intron, are known as the branch site, the polypyrimidine tract and, finally, the 3' (three-prime) splice site.

A separate regulatory system controls the splicing process by directing the basal machinery to these splice sites. More than 10 different splicing regulatory (SR) proteins have been identified. Their forms may vary in different tissues or stages of development in the same tissue. SR proteins can bind to short nucleotide sequences within the exons of the pre-mRNA transcript. These binding sites are known as exonic splicing enhancers (ESE) because when the appropriate SR protein binds to an ESE, that action recruits the basal machinery's snRNAs to the splice sites adjacent to either end of the exon. Yet an SR protein can also bind to an exonic splicing suppressor (ESS) sequence within the exon, which will suppress the basal machinery's ability to bind to the ends of that exon and result in its being spliced out of the final mRNA.

The effect of skipping just one exon can be dramatic for an organism. In fruit flies, for example, alternative splicing regulates the sex-determination pathway. When a gene called *Sex-lethal* is expressed, a male-specific exon may be skipped during splicing, leading to the synthesis of a female-specific Sex-lethal protein. This protein can then bind to any subsequent pre-mRNA transcripts from the same gene, ensuring that all further splicing events will continue to cut out the male-specific exon and guaranteeing that only the female-specific protein will be synthesized. If the male-specific exon is spliced in during the first round of editing, however, a nonfunc-

The effect of skipping just one exon can be dramatic for an organism.

tional mRNA results, which commits the fly's cells to the male-specific pathway.

Exon skipping is the most common type of alternative splicing found in mammals. But several other kinds have also been identified, including one that causes introns to be retained in mature mRNA, which is most prevalent in plants and lower multicellular lifeforms. Intron retention is probably the earliest version of alternative splicing to have evolved. Even today the splicing machinery of single-celled organisms, such as yeast, operates by recognizing introns, in contrast with the SR protein system of higher organisms, which defines exons for the basal machinery.

In the unicellular system the splicing machinery can recognize only intronic sequences of fewer than 500 nucleotides, which works fine for yeast because it has very few introns, averaging just 270 nucleotides long. But as genomes expanded during evolution, their intronic stretches multiplied and grew, and cellular splicing machinery was most likely forced to switch from a system that recognizes short intronic sequences within exons to one that recognizes short exons amid a sea of introns. The average human protein-coding gene, for example, is 28,000 nucleotides long, with 8.8 exons separated by 7.8 introns. The exons are relatively short, usually about 120 nucleotides, whereas the introns can range from 100 to 100,000 nucleotides long.

The size and quantity of human introns—we have the highest number of introns per gene of any organism—raises an interesting issue. Introns are an expensive habit for us to maintain. A large fraction of the energy we consume every day is devoted to the maintenance and repair of introns in their DNA form, transcribing the pre-mRNA and removing the introns, and even to the breakdown of introns at the end of the splicing reaction. Furthermore, this system can cause costly mistakes. Each miscut and ligation of pre-mRNA leads to a change in the gene transcript's protein-coding sequence and possibly to the synthesis of a defective protein.

For instance, an inherited disease that I am investigating, familial dysautonomia, results from a single-nucleotide mutation in a gene called *IKBKAP* that causes it to be alternatively spliced in nervous system tissues. The resulting decreased availability of the standard IKBKAP protein leads to abnormal development of the nervous system, and about half of all patients with this disease die before the age of 30. At least 15 percent of the gene mutations that produce genetic diseases (and probably certain cancers as well) do so by affecting pre-mRNA splicing. So why has evolution preserved such a complicated system that is capable of causing disease? Perhaps because the benefits outweigh the risks.

Advantages in Alternatives

BY GENERATING more than one type of mRNA molecule and, therefore, more than one protein per gene, alternative splicing certainly allows humans to manufacture more than 90,000 proteins without having to maintain 90,000 genes. On average, each of our genes generates about three alternatively spliced mRNAs. Still, that number does not explain our need for so many introns

THE AUTHOR

GIL AST is a senior lecturer in the department of human genetics and molecular medicine at the Tel Aviv University Medical School in Israel. His research focuses on the molecular mechanics of pre-mRNA splicing, the evolution and regulation of alternative splicing, and splicing defects associated with cancers and inherited diseases. He recently collaborated with scientists at Compugen to develop a bioinformatics system for predicting alternative-splicing events to detect novel protein variants.

and why they occupy the vast majority of real estate within genes, leaving exonic sequences to make up only 1 to 2 percent of the human genome.

After the sequencing teams had revealed this seemingly empty genomic landscape in 2001, yet another enigma arose when the mouse genome was published in 2002. It turned out that a mouse possesses almost the same number of genes as a human. Although approximately 100 million years have passed since we had a common ancestor, the vast majority of human and mouse genes derive from that ancestor. Most of these share the same intron and exon arrangement, and the nucleotide sequences within their exons are also conserved to a high degree. So the question becomes, if so little differs between the genomes of humans and mice, what makes us so vastly different from the rodents?

Christopher J. Lee and Barmak Modrek of U.C.L.A. recently revealed that one quarter of the alternatively spliced exons in both genomes are specific either to human or mouse. Thus, these exons have the potential to create species-specific proteins that could be responsible for diversification between species. Indeed, one group of alternatively spliced exons is unique to primates (humans, apes and monkeys) and might have contributed to primates' divergence from other mammals. By studying the process whereby such an exon is born, we can begin to see the advantages of introns in general, and the energy we expend to sustain them seems justified.

These primate-specific exons derive from mobile genetic elements called Alus, which belong to a larger class of elements known as retrotransposons— short sequences of DNA whose function seems to be generating copies of themselves and then reinserting those copies back into the genome at random positions, rather like little genomic parasites. Retrotransposons are found in almost all genomes, and they have had a profound influence by contributing to the genomic expansion that accompanied the evolution of multicellular organisms. Almost half the human genome is made

CHIMPANZEES AND HUMANS share 99 percent of their genomes, including tiny mobile genetic elements, called Alus, found only in primates. Alus may have given rise, through alternative splicing, to new proteins that drove primates' divergence from other mammals. Humans' divergence from other primates may also be thanks in part to alternative splicing: recent studies have shown that the nearly identical genes of humans and chimps produce essentially the same proteins in most tissues, except in parts of the brain, where certain human genes are more active and others generate significantly different proteins through alternative splicing of gene transcripts.

up of transposable elements, Alus being the most abundant.

Alu elements are only 300 nucleotides long with a distinctive sequence that ends in a "poly-A tail." Our genome already contains some 1.4 million Alu copies, and many of these Alu elements are continuing to multiply and insert themselves in new locations in the genome at a rate of about one new insertion per every 100 to 200 human births.

The Alus were long considered nothing more than genomic garbage, but they began to get a little respect as geneticists realized how Alu insertion can expand a gene's protein-generating capacity. About 5 percent of alternatively spliced exons in the human genome contain an Alu sequence. These exons most likely originated when an Alu element "jumped" into an intron of a gene, where the insertion normally would not have any negative consequence for the primate because most introns are spliced out and discarded. Through subsequent mutation, however, the Alu could turn the intron in which it resides into a meaningful sequence of genetic information—

an exon. This can happen if changes in the Alu sequence create a new 5′ or 3′ splice site within the intron, causing part of the intron to be recognized as "exon" by the spliceosome. (Such mutations usually arise during cell division, when the genome is copied and a "typo" is introduced.)

If the new Alu exon is only alternatively spliced in, the organism can enjoy the best of two worlds. By including the Alu exon, its cells can produce a novel protein. But the new capability does not interfere with the gene's original function, because the old types of mRNA are also still synthesized when the Alu exon is spliced out. Only when a mutated Alu becomes spliced constitutively—that is, the Alu exon is always spliced into all the mRNAs produced from the gene—does it become problematic, because it can trigger genetic diseases caused by the absence of the old protein. To date, three such genetic illnesses caused by misplaced Alu sequences have been identified: Alport and Sly syndromes and OAT deficiency.

My colleagues and I have shown

that all it takes to convert some silent intronic Alu elements into real exons is a single-letter change in their DNA sequence. At present, the human genome contains approximately 500,000 Alu elements located within introns, and 25,000 of those could become new exons by undergoing this single-point mutation. Thus, Alu sequences have the potential to continue to greatly enrich the stock of meaningful genetic information available for producing new human proteins.

RNA Therapy

MORE THAN 400 research laboratories and some 3,000 scientists worldwide are trying to understand the very complex reactions involved in alternative splicing. Although this research is still at a very early stage, these investigators agree that recent findings point toward future therapeutic applications, such as new gene therapy strategies that exploit the splicing mechanism to treat both inherited and acquired disorders, such as cancer.

One approach might be to direct a short stretch of synthetic RNA or DNA nucleotides, called antisense oligonucleotides, to bind to a specific target on the patient's DNA or RNA. Antisense oligonucleotides could be delivered into cells to mask either a specific splice site or some other regulatory sequence, thereby shifting the splicing activity to another site. Ryszard Kole of the University of North Carolina at Chapel Hill first demonstrated this technique on human blood progenitor cells from patients with an inherited disorder called beta-thalassemia, in which an aberrant 5' splice site causes oxygen-carrying hemoglobin molecules to be deformed. By masking the mutation, Kole was able to shift splicing back to the normal splice site and restore production of functional hemoglobin.

Later, Kole showed that the same technique could be used on human cancer cells grown in culture. By masking a 5' splice site of the Bcl-x apoptosis-regulating gene transcript, he was able to shift splicing activity to generate the Bcl-x(S) form of mRNA rather than the Bcl-x(L) form, decreasing the cancer cells' synthe-

sis of the antiapoptotic protein and enhancing synthesis of the proapoptotic protein. In some cancer cells, this change activates the apoptotic program; in others, it enhances the apoptotic effects of chemotherapeutic drugs administered along with the oligonucleotides.

Another way to use the alternative splicing mechanism for therapy was demonstrated in 2003 by Adrian Krainer and Luca Cartegni of Cold Spring Harbor Laboratory in Long Island, N.Y., who found a way to induce cells to splice in an exon that would otherwise be skipped. They created a synthetic molecule that can be programmed to bind to any piece of RNA according to its sequence, then attached the RNA-binding part of an SR protein. This chimeric molecule can therefore both bind to a specified sequence on the pre-mRNA and recruit the basal machinery to the appropriate splice signal. Krainer and Cartegni used this method on human cells grown in culture to correct splicing defects in mutated versions of the BRCA1 gene, which has been implicated in breast cancer, and of the SMN2 gene, which causes spinal muscular atrophy.

Yet a third approach capitalizes on the ability of the spliceosome to join two different pre-mRNA molecules from the same gene to form a composite mRNA. Termed trans-splicing, this event is common in worms but occurs only rarely in human cells. Forcing the spliceosome to

trans-splice could allow a mutated region of pre-mRNA responsible for disease to be precisely excised and replaced with a normal protein-coding sequence. Recently John Englehardt of the University of Iowa used this technique in cell culture to partially correct the pre-mRNA of a gene that produces a defective protein in the airway cells of cystic fibrosis sufferers.

Before the human genome was decoded, few scientists believed that organisms as complex as humans could survive with a mere 25,000 genes. Since the sequence was completed, alternative splicing has emerged as the pivotal process that permits a small number of genes to generate the much larger assortment of proteins needed to produce the human body and mind while precisely orchestrating their manufacture in different tissues at different times. Moreover, splicing explains how the tremendous diversity among humans, mice and presumably all mammals could originate in such similar genomes.

Evolution works by presenting organisms with new options, then selecting to keep those that confer an advantage. Thus, novel proteins created by the splicing in of new Alu-derived exons probably helped to make humans the species we are today. And further investigation of the alternative splicing process promises still greater improvements in our quality of life.   ⬛

MORE TO EXPLORE

Alternative Splicing: Increasing Diversity in the Proteomic World. B. R. Graveley in *Trends in Genetics*, Vol. 17, No. 2, pages 100–107; February 2001.

Splicing Regulation as a Potential Genetic Modifier. M. Nissim-Rafinia and B. Kerem in *Trends in Genetics*, Vol. 18, No. 3, pages 123–127; March 2002.

The Hidden Genetic Program of Complex Organisms. John S. Mattick in *Scientific American*, Vol. 291, No. 4, pages 60–67; October 2004.

How Did Alternative Splicing Evolve? Gil Ast in *Nature Reviews Genetics*, Vol. 5, pages 773–782; October 2004.

Why has evolution preserved a complicated system that can cause disease?

Article Review: CELL BIOLOGY

The Alternative Genome *by Gil Ast*

TESTING YOUR COMPREHENSION

1. Of what macromolecular component(s) is a splicesome made? How does a splicesome compare with a ribozyme with regard to composition and function?

    ANSWER: *The splicesome is composed of RNA and protein. The RNA components are named U1, U2, U4, U5 and U6, while there are many small proteins in the splicesome complex. Ribozymes are broadly defined as catalytically active RNAs. These RNAs can splice introns from their own RNA structure. The rRNA in the ribosome catalyzes the formation of peptide bonds during translation, and thus is considered a type of ribozyme.*

2. Define the cellular process of identifying an exon, and defining the exon/intron margins for the exon. Identify the components of this sequential process.

    ANSWER: *Splicing regulatory (SR) proteins bind to an RNA sequence within the exon. This RNA sequence is called the exon splicing enhancer (ESE) site. The binding of SR proteins to the RNA helps to recruit small RNAs (U1 and U2). These small RNAs bind to the transcript specifically at the splicing sites, and their presence is key to splicesome formation and activity at that site.*

3. What is one known mechanism for identifying an exon and suppressing its inclusion in the final RNA? What protein(s) mediate this process?

    ANSWER: *The splicing regulatory (SR) protein, in addition to identifying exons for inclusion in the final mRNA, can also identify exons for exclusion from the final mRNA. Thus the SR protein can serve as a splicing suppressor, binding to exonic splicing suppressor (ESS) sites. There are multiple SR proteins that have been characterized, they differentially regulate splicing depending upon the cell type or stage of differentiation.*

CLASS ACTIVITIES AND DISCUSSION

1. Apoptosis, or programmed cell death, is a carefully regulated cellular program that can be activated to eliminate damaged, infected, or unwanted cells within a multi-cellular organism. The splicing of the Bcl-x mRNA has been shown to dictate whether a small protein (S) or a large protein (L) is produced. Bcl-x (L) protein is an apoptosis suppressor while the Bcl-x (S) protein activates apoptosis. What cellular conditions or exposures do you predict to influence the splicing of Bcl-x transcripts? Discuss the process of apoptosis, and the regulation of apoptosis by Bcl-x (L)/(S). Upon examination of the literature (PubMed - try "regulation of Bcl-x splicing in neurons"), what conditions were identified?

    Original Paper:
    Boise et al. (1993), bcl-x, a bcl-2-related gene that functions as a dominant regulator of apoptotic cell death, *Cell* 74(4):597-608.

2. A central example of alternative splicing in cell biology is within the mechanism of sex determination in fruit flies (*Drosophila melanogaster*). There is a cascade of splicing regulation events in three separate genes, culminating with the generation of a male or female-specific protein from the doublesex (dsx) gene, depending upon an alternative splicing choice. Divide the class into three groups, each responsible for a separate gene/protein: Sex-lethal (sxl), transformer (tra), or doublesex (dsx). Each of these genes has been well characterized in fruit flies. Excellent sources of information will include textbooks, review articles, and research articles. As a class, work together to outline the sex determination pathway, and the role alternative splicing plays in that process. A key point to notice is the dual capabilities of doublesex, and the ability of alternative splicing controlling this gene's impact upon sex determination.

Article Review: GENETICS

The Alternative Genome *by Gil Ast*

TESTING YOUR COMPREHENSION

1. In anticipation of completion of the human genome project, scientists around the world speculated on the number of genes contained within the human genome. How many genes did most researchers expect? Describe their reasoning for making these predictions.

   ANSWER: As the human genome was being completed, the genomes of simpler organisms were completed and shared with the public. The numbers of genes seemed to increase with the complexity of the organism. The yeast Saccharomyces cerevisiae was shown to have ~ 6,500 genes, and the roundworm Caenorhabditis elegans was shown to have ~19,500 genes. In addition, experimental data had predicted that humans generate ~90,000 different proteins, thus it was predicted that we would have close to 90,000 genes.

2. Alternative splicing can be categorized in a number of different ways beyond simply skipping certain exons when generating the final mRNA. Define those categories, and diagram an example gene splice for each category.

   ANSWER: As is illustrated in the first figure, alternative splicing can occur via skipping exons, use of alternative 5' or 3' splice sites within an intron, the retention of an exon in the final mRNA, and/or the use of mutually exclusive exon retention strategies (one exon retained from a group of exons).

3. Define an Alu sequence. What relevance does this type of sequence have to a discussion of alternative splicing?

   ANSWER: An Alu sequence is a type of retro-transposon, mobile genetic elements that are found throughout the human genome (called jumping genes or a genetic parasites). The human genome has 1.4 million Alu sequences (40 times the total number of genes). As these sequences move about the genome, they can insert themselves into exons and introns, generating genetic changes that sometimes generate new exons. A single mutation within an Alu sequence is sufficient to cause recognition of that sequence as an exon.

CLASS ACTIVITIES AND DISCUSSION

1. Researchers are beginning to examine the possible manipulation of alternative splicing as a form of gene therapy, using the tools of molecular biology to control the expression of genes that are responsible for disease. Divide the class into three groups, each being responsible for a careful examination of one gene therapy strategy that employs the manipulation of gene splicing. Students within each group will work together to develop a class presentation on their topic. Each group will identify key researchers in their assigned field of work, they will define the gene therapy strategy and the underlying cell biology, and the will describe the disease(s) that may be treated with this therapy. The three categories are defined in the "RNA Therapy" section of the manuscript, and include: 1) antisense oligonucleotides to affect the splicing of a selected gene, 2) synthetic splice regulators, and 3) the activation of controlled trans-splicing" in human cells.
   Interesting Website: ExonHit Therapeutics (alternative splicing and gene therapy) www.exonhit.com

2. Famililial dysautonomia is caused by mutation of a gene known as "inhibitor of kappa light polypeptide gene enhancer in B-cells, kinase complex-associated gene" – mercifully shortened to IKBKAP or IKAP. The author mentions his research work characterizing this disease. The mutation causes the abnormal splicing of the IKAP transcript in neuronal cells, affecting neuronal development and function in affected individuals.

   From what you've learned in this paper, discuss this disease in the context of a neuron-specific affect. If the mutation affects IKAP splicing, why aren't all the cells of the body affected by this mutation? For additional help, use the National Library of Medicine to link into the Genetics Home Reference site (http://ghr.nlm.nih.gov/).